Neurocomputation in Remote Sensing Data Analysis

Springer-Verlag Berlin Heidelberg GmbH

I. Kanellopoulos · G. G. Wilkinson
F. Roli · J. Austin (Eds.)

Neurocomputation in Remote Sensing Data Analysis

Proceedings of Concerted Action COMPARES
(Connectionist Methods for Pre-Processing
and Analysis of Remote Sensing Data)

With 87 Figures
and 39 Tables

Springer

Ioannis Kanellopoulos
Joint Research Centre, Commission of the European Communities
Space Applications Institute
Environmental Mapping and Modelling Unit
I-21020 Ispra (Varese), Italy

Prof. Graeme G. Wilkinson
Kingston University
Head, School of Computer Science and Electronic Systems
Penrhyn Road
Kingston Upon Thames, KT1 2EE, United Kingdom

Dr. Fabio Roli
University of Cagliari
Electrical and Electronic Engineering Department
Piazza d'Armi
I-09123 Cagliari, Italy

Dr. James Austin
University of York, Department of Computer Science
Advanced Computer Architecture Group
York, Y01 5DD, United Kingdom

ISBN 978-3-642-63828-2

Cataloging-in-Publication Data applied for
Die Deutsche Bibliothek - CIP-Einheitsaufnahme
Neurocomputation in remote sensing data analysis : proceedings of concerted
action COMPARES (connectionist methods for pre-processing and analysis of
remote sensing data) ; with 39 tables / I. Kanellopoulos ... (ed.). - Berlin ;
Heidelberg ; New York ; Barcelona ; Budapest ; Hong Kong ; London ; Milan ;
Paris ; Santa Clara ; Singapore ; Tokyo : Springer, 1997
   ISBN 978-3-642-63828-2   ISBN 978-3-642-59041-2 (eBook)
   DOI 10.1007/978-3-642-59041-2

© Springer-Verlag Berlin Heidelberg 1997
Originally published by Springer-Verlag Berlin · Heidelberg New York in 1997

The use of general descriptive names, registered names, trademarks, etc. in this publication does not imply, even in the absence of a specific statement, that such names
are exempt from the relevant protective laws and regulations and therefore free for
general use.

Product liability: The publishers cannot guarantee the accuracy of any information
about the application of operative techniques and medications contained in this
book. In every individual case the user must check such information by consulting
the relevant literature.

Hardcover-Design: Erich Kirchner, Heidelberg

SPIN 10636065   42/2202-5 4 3 2 1 0 – Printed on acid-free paper

Foreword

Since 1994 the European Commission has been supporting activities under the Environment and Climate programme of research and technological development, with the aim of developing cost-effective applications of satellite Earth observation (EO) for both environmental monitoring and research.

This action has included support to methodological research, aimed at the development and evaluation of new techniques forming part of the chain of processing needed to transform data into useful information. Wherever appropriate, the Commission has emphasised the coordination of ongoing research funded at the national level, through the mechanism of concerted actions. Concerted actions are flexible and efficient means to marshal efforts at the European level for a certain period. They are proposed by groups of researchers active in a given field who have identified the added value to be gained by European cooperation, whilst continuing to pursue their own individual projects.

In view of the rapid developments in the field of neural network over the last 10 years, together with the growing interest of the Earth observation community in this approach as a tool for data interpretation, the Commission decided in 1995 to support the concerted action COMPARES, following a proposal from a group of acknowledged European experts.

The COMPARES concerted action provided a forum for a thorough review of the state-of-the-art, and a Europe-wide debate on future research priorities. The present volume, arising from a highly successful workshop held in July 1996, will provide a valuable reference for those working in this field, and will help indicate the most promising applications of neural networks for further development and eventual operational use.

Alan Cross
Environment & Climate Programme
DG XII: Science, research and development
European Commission

Table of Contents

Foreword V

Introduction 1

Open Questions in Neurocomputing for Earth Observation
Graeme G. Wilkinson 3

A Comparison of the Characterisation of Agricultural Land
Using Singular Value Decomposition and Neural Networks
S. Danaher, G. Herries, T. Selige, and M. Mac Súirtán 14

Land Cover Mapping from Remotely Sensed Data with a
Neural Network: Accommodating Fuzziness
Giles M. Foody 28

Geological Mapping Using Multi-Sensor Data: A
Comparison of Methods
Paul M. Mather, Brandt Tso, and Magaly Koch 38

Application of Neural Networks and Order Statistics Filters
to Speckle Noise Reduction in Remote Sensing Imaging
E. Kofidis, S. Theodoridis, C. Kotropoulos, and I. Pitas 47

Neural Nets and Multichannel Image Processing
Applications
Vassilis Gaganis, Michael Zervakis, and Manolis Christodoulou........ 57

Neural Networks for Classification of Ice Type
Concentration from ERS-1 SAR Images. Classical Methods
versus Neural Networks
Jan Depenau 71

A Neural Network Approach to Spectral Mixture Analysis
Theo E. Schouten and Maurice S. klein Gebbinck 79

Comparison Between Systems of Image Interpretation
Jean-Pierre Cocquerez, Sylvie Philipp, and Philippe Gaussier 86

Feature Extraction for Neural Network Classifiers
Jon A. Benediktsson and Johannes R. Sveinsson 97

**Spectral Pattern Recognition by a Two-Layer Perceptron:
Effects of Training Set Size**
Petra Staufer and Manfred M. Fischer 105

**Comparison and Combination of Statistical and Neural
Network Algorithms for Remote-Sensing Image
Classification**
Fabio Roli, Giorgio Giacinto, and Gianni Vernazza 117

**Integrating the Alisa Classifier with Knowledge-Based
Methods for Cadastral-Map Interpretation**
Eleni Stroulia and Rudolf Kober 125

**A Hybrid Method for Preprocessing and Classification of
SPOT Images**
Shan Yu and Konrad Weigl 134

**Testing some Connectionist Approaches for Thematic
Mapping of Rural Areas**
Leopoldo Cortez, Fernando Durão, and Vitorino Ramos 142

**Using Artificial Recurrent Neural Nets to Identify Spectral
and Spatial Patterns for Satellite Imagery Classification of
Urban Areas**
Sara Silva and Mario Caetano 151

**Dynamic Segmentation of Satellite Images Using Pulsed
Coupled Neural Networks**
X. Clastres, M. Samuelides, and G. L. Tarr....................... 160

**Non-Linear Diffusion as a Neuron-Like Paradigm for
Low-Level Vision**
M. Proesmans, L. J. Van Gool, and P. Vanroose 168

**Application of the Constructive Mikado-Algorithm on
Remotely Sensed Data**
C. Cruse, S. Leppelmann, A. Burwick, and M. Bode 176

A Simple Neural Network Contextual Classifier
Jens Tidemann and Allan Aasbjerg Nielsen 186

Optimising Neural Networks for Land Use Classification
Horst Bischof and Aleš Leonardis 194

High Speed Image Segmentation Using a Binary Neural Network.
Jim Austin 202

Efficient Processing and Analysis of Images Using Neural Networks
Stefanos Kollias 214

Selection of the Number of Clusters in Remote Sensing Images by Means of Neural Networks
Paolo Gamba, Andrea Marazzi, and Alessandro Mecocci 224

A Comparative Study of Topological Feature Maps Versus Conventional Clustering for (Multi-Spectral) Scene Identification in METEOSAT Imagery
P. Boekaerts, E. Nyssen, and J. Cornelis 232

Self Organised Maps: the Combined Utilisation of Feature and Novelty Detectors
C. N. Stefanidis and A. P. Cracknell 242

Generalisation of Neural Network Based Segmentation Results for Classification Purposes
Ari Visa and Markus Peura 255

Remote Sensing Applications Which may be Addressed by Neural Networks Using Parallel Processing Technology
Charles Day 262

General Discussion
Graeme G. Wilkinson 281

Introduction

Neural network or connectionist algorithms have made an enormous impact in the field of signal processing over the last decade. Although few would regard them as the perfect answer to all pattern recognition problems, there is little doubt that they have contributed significantly to the solution of some of the most difficult ones. Trainable networks of primitive processing elements have been shown to be capable of describing and modelling systems of great complexity without the necessity of building parameterised statistical descriptions. Such capability has led to considerable and constantly growing interest from the remote sensing community. From early beginnings in the late 1980's, neural network algorithms are now being explored for a wide range of uses in Earth observation. In many cases these uses are still experimental or at the pre-operational stage.

There is still a need to share experiences and to assess the best approaches to solve particular kinds of pattern recognition problems in remote sensing. Moreover with each new generation of satellites, the complexity and capture rate of remotely sensed data grow considerably making it essential to develop new product extraction procedures which are powerful, yet based on simple mathematical approaches, and are also suitable for high speed distributed or parallel forms of computer architectures. It is no surprise therefore that much attention has recently focused on neural network or connectionist approaches.

This volume contains chapters based on presentations made at the York workshop which was held as part of the one year European Concerted Action "COMPARES" (*COnnectionist Methods for Pre-processing and Analysis of REmote Sensing data*) during July 1996. The objective of this workshop was to bring together key researchers -primarily but not exclusively from within the European Union- to expose and debate some of the potential connectionist approaches which can be used in remote sensing and to highlight issues and topics for future research. Not unsurprisingly, it became apparent both from the presentations made in York and from the chapters prepared for this volume that a very wide range of different neural network models are now being explored in remote sensing and that they are being used for a considerable range of applications. During the workshop, the sessions were focused on particular themes such as the applications of multilayer neural networks, data selection and representation, hybrid methods for data analysis, alternative neural network methods, performance, design and implementation issues, the applications of neural data clustering methods, and finally future prospects. In general there emerged a significant amount of optimism at the York meeting, and an appreciation that collectively the application of connectionist systems in remote sensing is moving forward at a fast pace, in

Europe, with many interesting results. We believe this optimism is reflected in the individual chapters within this present volume.

Whilst it would not be appropriate to attempt to summarise the whole book here, one key finding that emerged from the York workshop was that although multispectral image classification has dominated so far as the primary application of neural networks in remote sensing, alternative uses such as signal inversion appear to be equally important. In the classification context, neural networks appear to gain over more traditional statistical methods because they do not require the data set to conform to a fixed statistical model. This aspect becomes very powerful in the context of fusion of multiple and independently gathered data sets -an increasing requirement in Earth observation. Also in the signal inversion area, neural networks appear to gain over more traditional approaches, but in this case because they can be used to describe a non-linear mapping from one data space to another. This non-linearity aspect also appears to be one of the main strengths of some connectionist approaches. Although the York workshop provided a forum for the exchange of ideas and views on many such issues, the audience was necessarily limited in number and one of the aims of this book is to bring some of the topics to a wider audience in the hope that it will lead to an acceleration of the pace of discovery in the field of connectionism applied to Earth observation. There is much work still to be done, though we can confidently say that the foundations of a whole new generation of algorithms for the analysis of remote sensing data have already been laid. It remains to be seen if the connectionist approaches described herein really begin to take over as the basis of most core analytical procedures in Earth observation in the years to come.

The editors would like to acknowledge the high quality of presentations made at the York workshop and to thank the authors of the chapters included here, for their cooperation in making what we hope will be an important contribution to the science of remote sensing. We would also like to thank Ken Lees in the Department of Computer Science at the University of York for the excellent organisation of the scientific and social events during the workshop. We are also grateful to Charles Day and Susan Harding for their assistance in the editing of the proceedings.

Finally we would like to express our gratitude to the Environment and Climate research and development programme of the European Commission for supporting the COMPARES initiative (contract no. ERB ENV4–CT95–0151).

Ioannis Kanellopoulos
Graeme G. Wilkinson
Fabio Roli
James Austin

J.R.C. Ispra
February 1997

Open Questions in Neurocomputing for Earth Observation

Graeme G. Wilkinson

School of Computer Science and Electronic Systems,Kingston University,
Penrhyn Road, Kingston Upon Thames, Surrey KT1 2EE, UK.

Summary. Neural network usage in remote sensing has grown dramatically in
recent years - mostly for classification. However neural systems are susceptible to
problems such as unpredictability, over-fitting and chaos which render them unsatisfying for many potential users. In the classification context they have yet to show
any major improvement over conventional statistical algorithms in generalisation
to large geographical areas (regional - continental scale). Apart from classification,
neural computation can solve a wider range of problems in remote sensing than
have been explored so far -especially related to geometrical processing and signal
inversion. Further important issues are raised which should form the basis for future
research such as the need for very large networks (which are sensitive to geographical and temporal context), the requirement for special purpose hardware, and the
need for better user-adapted systems.

1. Introduction: The Growth and Uses of Neural Computing in Remote Sensing

Since the early 1980's there has been a steady global growth in the use of
neural network algorithms for pattern recognition problems in diverse fields.
The field of remote sensing is no exception to this and neural networks began
to be used in the pre-processing and analysis of satellite data around 1988.
There has been an almost exponential growth since then as indicated by
the recent trend in numbers of publications which refer to the use of neural
methods in remote sensing.

Overall, most of the effort devoted to the use of neural networks in remote sensing has been oriented towards the classification problem, i.e. the
transformation of spectral radiances measured on pixels into thematic map
classes which describe landscapes in cartographic terms (e.g. [13]–[11]). This
classification problem is typical of the vast majority of pattern recognition
problems which are experienced in a wide variety of fields. As in other fields,
several different types of neural network models have been used in the classification of remotely sensed data (e.g. multi-layer perceptrons (MLP), self-
organising maps, counter-propagation nets, fuzzy adaptive resonance theory
(ART) nets...). Also some hybrid methods have been developed which either
integrate different neural models (e.g. MLP and learning vector quantisation
(LVQ) -[9]- or integrate neural models (such as MLP) with conventional statistical approaches such as the maximum-likelihood classifier (MLC) [22].
Above all, the interest in using neural network methods has been motivated

by their suitability for problems related to mixed data sets which do not follow a known statistical distribution which could be parameterised. Hence neural network methods have found favour over approaches such as MLC in the analysis of mixed source data sets (e.g. [24] - [12]) and mixed time frame data sets (e.g. [10]).

Apart from the overwhelming interest in the use of neural networks for classification in remote sensing, there has been a small but growing awareness that they can also be applied to more general problems involving transformations from one data space to another e.g. in the retrieval of complex surface parameters by inversion of satellite measurements [18]. Also, whilst this is still a largely undeveloped area, neural networks are beginning to find uses in geometrical operations on imagery such as geometrical rectification [15], linear object detection [8], [20], and stereo-image matching [16].

Apart from their generic uses, neural networks have been employed in remote sensing for a wide range of environmental or thematic applications such as: forest mapping [23]; crop mapping [6]; rainfall and precipitation cell-top estimation [25], [21]; land cover mapping [2] and land cover change detection [7]; and sea ice mapping [14], [17].

2. Problems in Using Neural Networks

Despite the growing interest in the use of neural networks in remote sensing, a number of difficulties are associated with their use which restricts their general acceptability in geographical data analysis.

One of the most difficult aspects of using neural networks is that they rely on simple training algorithms such as "back-propagation" [19] which involve a search for a network weight combination in a mathematical space of high dimensions to give a global minimum in pattern classification error. These algorithms are not guaranteed to find the solution giving the global error minimum in less than infinite time and the search procedure is unpredictable and may fall into local minima (figure 2.1) Although numerical techniques such as "simulated annealing" or use of genetic algorithm approaches can improve search strategies, the problem of finding the global error minimum in less than infinite time can not be guaranteed. For this reason users have to accept the best result acquired in a realistic time scale -even if it is not optimum. The level of accuracy which can be attained is usually totally unpredictable at the start of a training sequence and can also vary significantly depending on starting conditions for the training (i.e. network architecture or configuration, initial random weight settings, training algorithm control parameter settings). This general unpredictability of behaviour can be regarded as a drawback for many potential users of neural systems.

Another critical issue in the training of neural networks is the "overfitting" problem. Overfitting occurs when a network adapts its internal weights too closely to the characteristics of the limited set of data with which

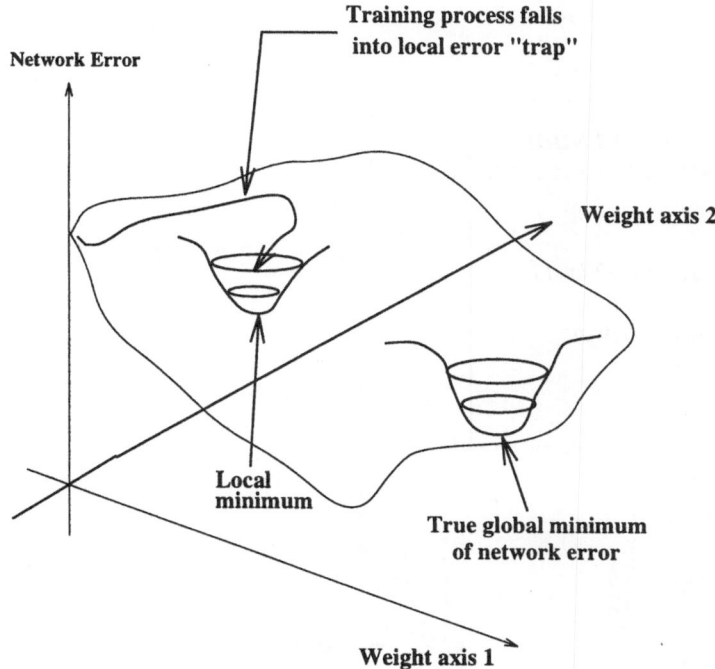

Fig. 2.1. Collapse of training into local error minimum

it has been trained. Although adapting to training data is essentially desired behaviour for a neural network, if this goes too far the network may fit too closely to the training data and perform poorly in recognition tasks on unseen data. This problem is not easily avoided and the only realistic approach is to check recognition performance on "un-seen" data at regular intervals as training proceeds and if the recognition performance begins to degrade to stop the training. In general it is difficult to know precisely at what point good training and adaptive behaviour of a network degrades into over-fitting. This is another aspect which for many potential users can be regarded as unsatisfying.

An altogether different, and less well-known problem associated with neural networks, is their susceptibility to chaotic behaviour. Most neural networks have non-linear transfer functions at each node which can, in the right combination of circumstances, lead to chaotic behaviour in training [3]. Chaotic regimes are associated with non-linear phenomena and neural networks are no exception to this. This is often manifested as extreme sensitivity to arithmetic rounding errors during training (figure 2.2). Nevertheless, despite this potential draw-back the non-linear aspect of a neural system is one of its most powerful features which allows it to solve complex pattern recognition problems.

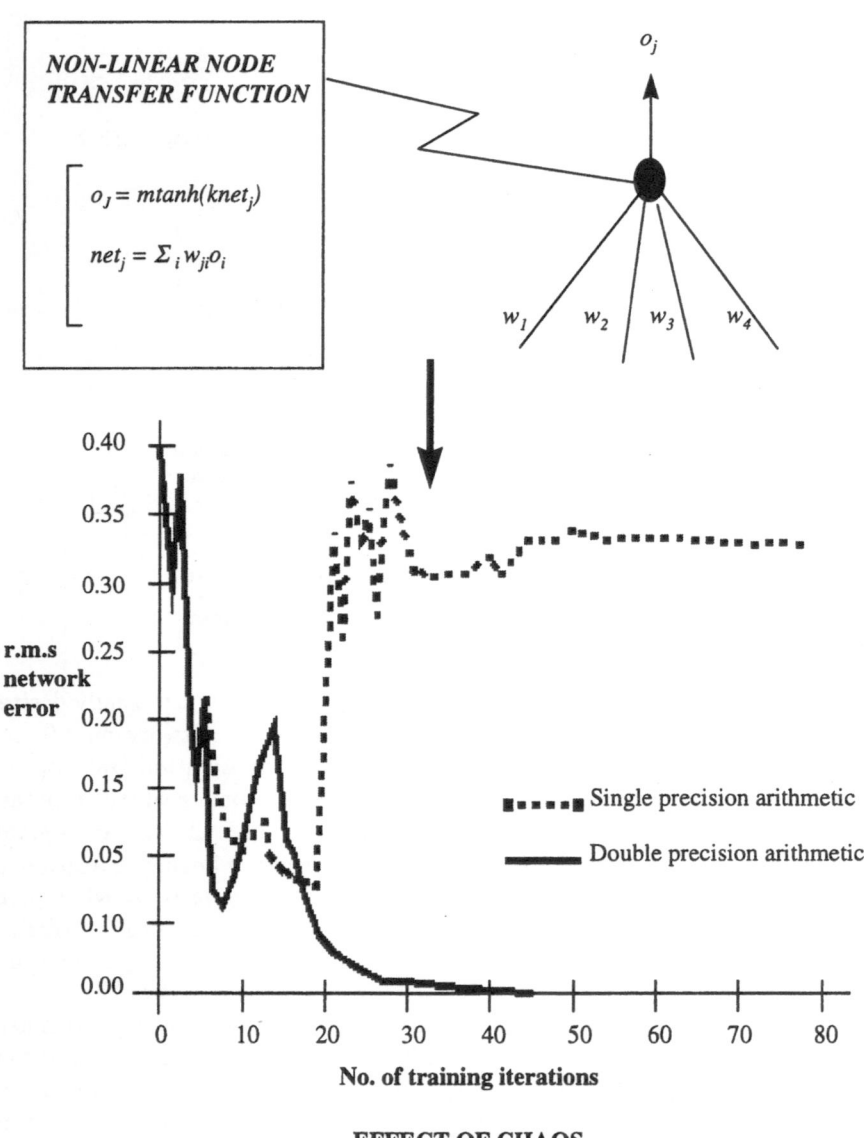

Fig. 2.2. Chaos manifested in network training as over-sensitivity to arithmetic rounding error.

3. Issues Related to Image Classification

Classification has remained the pre-dominant use of neural networks in remote sensing over the last few years and this continues to be the area in which their use is most understood. However, recent experience has shown, that despite a significant level of effort to improve classification performance through use of neural networks, no noticeable improvement in accuracy has been achieved in remote sensing to date through the employment of the neural approaches. It is not clear why this is the case, though a few clues can be gathered.

A contributory factor in the lack of progress in classification by use of neural algorithms is almost certainly the relatively poor quality of ground "truth" data which are often used in training. The aphorism "garbage in, garbage out" applies to neural networks as much as it does to other forms of classifier. Vague statements have often been made about the robustness of neural classifiers in the presence of noise, at least in comparison to statistical classifiers, though this claim remains to be demonstrated in the classification of remote sensing data.

A significant problem encountered in all attempts to classify remote sensing data is that classifiers trained in one locality rarely work well when data from a different locality are used -the geographical generalisation problem. The experience to date with conventional classifiers is that they do not generalise well to different or unfamiliar geographical locations. Recent experiments have also shown that neural network classifiers are prone to precisely the same problem. Table 3.1 shows total classification accuracy obtained with classifiers trained on data from one locality and then used to classify imagery from a different (foreign) locality with and without some local re-training. The results are unequivocal in demonstrating that even when the same type of data inputs are used and the same classes are to be extracted, a neural network classifier trained in one geographical region will not "port" to another. This is fundamentally unsatisfactory for remote sensing in which the main benefit of the technology is the ability to observe and categorise extensive territorial areas at an economic cost.

One potential solution to this is to consider the creation of a single very large neural network (VLNN) which has sufficient flexibility and encoding power to describe the characteristics of landscapes in different geographical regions. Such a classifier may be a one-off product which would serve for many different mapping projects. The creation of such a VLNN -which could become, for example, a "pan-European classifier"- would be a formidable task (figure 3.1).

Such a VLNN would almost certainly need to be developed using special purpose neural hardware systems which exploit parallelism such as the SYNAPSE-1 Neurocomputer developed by Siemens-Nixdorf (see article by Day in this volume). Another approach towards improving classification performance could involve the use of fuzzy landscape categories instead of fixed

Table 3.1. Test of geographical generalisation ability of neural network classifier. Image data (Landsat TM and SPOT-HRV) plus ground truth data were taken from two diverse geographical locations (Lisbon, Portugal and Ardèche, France). Classification was performed using three spectral and three texture features from the imagery. Nine land cover classes were extracted (Adapted from [4]).

	Nature of Experiment	Total % Correct Pixel Classification
"Local"	Classification of Lisbon area image using network trained with local ground truth	71.1%
"Local"	Classification of Ardèche area image using network trained with local ground truth	63.9%
"Foreign"	Classification of Lisbon area image using network trained with Ardèche area ground data	14.7%
"Foreign"	Classification of Ardèche area image using network trained with Lisbon area ground data	12.7%
"Foreign + Re-train"	Classification of Lisbon area image using network trained with Ardèche area ground data plus partial re-training with local ground data	67.1%
"Foreign + Re-train"	Classification of Ardche area image using network trained with Lisbon area ground data plus partial re-training with local ground data	51.2%

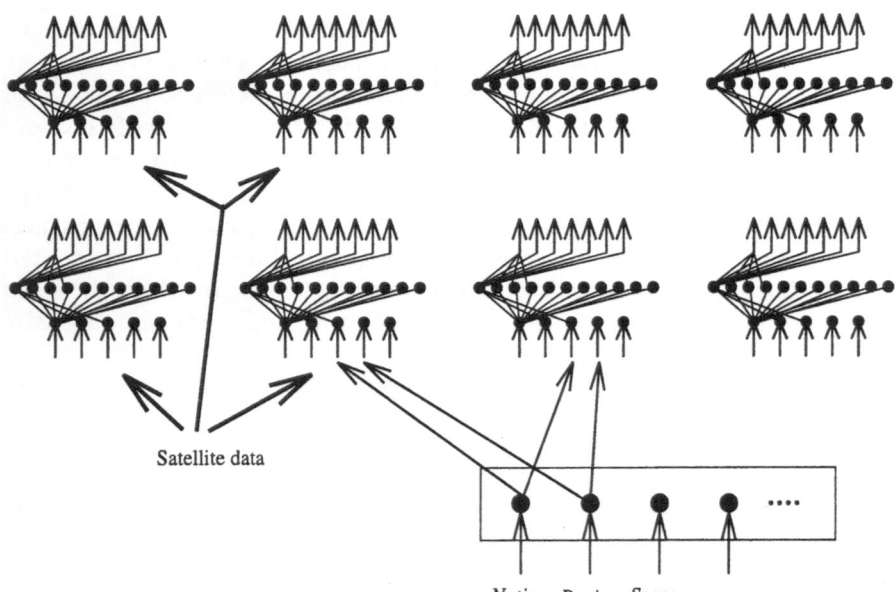

Satellite data

Nation Region Season ••••

Non-satellite data (geographic,temporal context etc.)

Fig. 3.1. The concept of "pan-European" network (VLNN) satellite image classifier which has separate sub-networks to process different geographical and temporal contexts.

classes. There has been growing interest in this topic in recent years and some initial results have confirmed the possibility of using neural networks to generate class mixture components from satellite data [5] - [1].

4. Potential New Applications of Neural Computing in Remote Sensing

Although, as stated earlier, there has been an almost exponential growth in the use of neural networks in remote sensing over the last seven years, this growth has mostly come about through wider use of existing neural network models for solving the same types of problem. A very small part of the growth in neural network use in remote sensing actually reflects real diversification of purpose for which the network is used. In general since neural networks perform general mathematical transformations

REPRESENTATION 1 → N.N. → REPRESENTATION 2

they could be used for a very wide range of data analysis tasks in remote sensing. The recent, but rare, use of neural networks for signal inversion is an interesting development which certainly should be followed up. However it is possible that new avenues should be considered, such as the possibility for use of neural networks in early processing of satellite data (calibration, geometrical correction -some work has been done on this-, atmospheric correction, multi-resolution signal merging, noise removal...).

Also neural networks may provide a suitable mechanism for integrating contextual information (e.g. from a GIS) in data analysis, though techniques are needed to handle combinations of ordinal and non-ordinal data sets for example, which could be an important topic for future research. Neural networks may also have roles in the selection of optimal band combinations or transformations when complex data sets are in use comprising a large number of multi-spectral or multi-temporal "channels". The possibility may also exist to utilise neural networks in the automatic selection and triggering of other algorithms to be applied to a particular data set, though at present this remains largely uncharted territory in the field of remote sensing.

5. Open Questions for Future Research

Overall it is clear that a number of issues related to the use of neural networks in remote sensing remain to be answered and further research studies are required to answer them. Among the most critical questions for further study are the following:

1. Do neural networks really offer significant advantages compared to other pattern recognition and data transformation algorithms ?
2. Classification has been the main application for neural networks in Earth observation but has research on classification reached an impasse imposed by extraneous factors such as quality of ground data, lack or impossibility of precise class definitions ?
3. Is it necessary or even possible to construct a very large modular neural network (VLNN) to encode landscape characteristics of the whole of Europe -i.e. to create a "pan-European classifier" which can describe local conditions and avoid the generalisation problem ?
4. If neural network algorithms are useful and could become part of the standard tool-box for environmental data analysis how can they be made totally user-friendly such that they can be used by environmental scientists with minimum knowledge of their inner functionality ?
5. Is special purpose hardware really needed to exploit neural networks in a realistic way in remote sensing in an operational context -e.g. is parallel hardware a pre-requisite to achieve satisfactory turn round times in image analysis ?
6. Should new or less common neural network models and architectures be explored for use in remote sensing or can the existing commonly-used models such as MLP offer as much functionality as is likely to be required for most practical purposes ?
7. Are there any novel applications of neural networks in remote sensing that have so far not been considered (such as modelling atmospheric corrections to raw satellite radiances) ?

Such questions should form the basis of future work in this field and are critical in assessing the long term role of connectionist computing in Earth observation. Some of the other articles in this volume partially cover some of these points, though the majority of these issues have still not been answered satisfactorily. Some views on these issues are given in the discussion section in this volume.

6. Discussion

In this article, we have tried to show that despite the rapid rise in popularity of neural networks for the analysis of remote sensing data, it is generally apparent that the data processing problems to which they are being applied (such as classification) are not currently being solved with any more reliability than with conventional algorithms. At the same time, however, it is apparent that neural networks potentially have a wider range of applications in remote sensing than just those for which they have been used to date and that they may be easier to scale to more complex problems than existing methods. Hence it is possible that they could become an optimal component of the remote sensing tool-box in the future. It is also apparent that on account of their generality, they could be used in an integrated way to solve multiple problems in remote sensing with only one underlying mathematical model. This could be an attractive feature of neural networks as it may lead to efficiency of computer code and reusable or modular multi-purpose data analysis systems. At the same time a number of key issues remain for future investigation such as the need for special purpose hardware.

Acknowledgement. The author is grateful to the following people at the Joint Research Centre of the European Commission, Ispra, who over several years contributed to the research reported in this article - especially Ioannis Kanellopoulos, Freddy Fierens, Zhengkai Liu, Alice Bernard, Walter Di Carlo, and Charles Day. The author would also like to acknowledge the support received from the European Commission's Environment and Climate research programme, whilst he was formerly at the Joint Research Centre, for the Concerted Action "COMPARES" (COnnectionist Methods for Pre-processing and Analysis of REmote Sensing data) –Contract no. ERB ENV4-CT95-0151.

References

1. A. C. Bernard, I. Kanellopoulos, and G. G. Wilkinson, "Neural network classification of mixtures", in *Soft Computing in Remote Sensing Data Analysis, Milano, Italy* (E. Binaghi, P. A. Brivio, and A. Rampini, eds.), vol. 1 of *Series in Remote Sensing*, pp. 53–58, Consiglio Nazionale Delle Ricerche, Italy, World Scientific, Singapore, December 1995.
2. D. L. Civco, "Artificial neural networks for land-cover classification and mapping", *International Journal of Geographical Information Systems*, vol. 7, no. 2, pp. 173–186, 1993.
3. H. L. J. Van der Maas, P. F. M. J. Verschure, and P. C. M. Molenaar, "A note on chaotic behaviour in simple neural networks", *Neural Networks*, no. 3, pp. 119–122, 1990.
4. DIBE - University of Genoa, "Portability of neural classifiers for large area land cover mapping by remote sensing", Report published under contract to the European Commission's Joint Research Centre, Ispra, Italy, 1995. Contract No: 10420-94-08 F1 ED ISP I.
5. G. M. Foody, "Approaches for the production and evaluation of fuzzy land cover classifications from remotely-sensed data", *International Journal of Remote Sensing*, vol. 17, no. 7, pp. 1317–1340, 1996.
6. G. M. Foody, M. B. McCulloch, and W. B. Yates, "Crop classification from C-band polarimetric radar data", *International Journal of Remote Sensing*, vol. 15, no. 14, pp. 2871–2885, 1994.
7. S. Gopal, D. M. Sklarew, and E. Lambin, "Fuzzy neural network classification of landcover change in the Sahel", in *Proceedings Eurostat/DOSES Workshop on " New tools for Spatial Data Analysis", Lisbon 18–20 November*, Eurostat, Luxembourg, November 1993.
8. O. Hellwich, "Detection of linear objects in ERS-1 SAR images using neural network technology", in *Proceedings of the International Geoscience and Remote Sensing Symposium (IGARSS'94), Pasadena, California*, vol. IV, pp. 1886–1888, IEEE Press, Piscataway, NJ, August 1994.

9. R. Hernandez, A. Varfis, I. Kanellopoulos, and G. G. Wilkinson, "Development of MLP/LVQ Hybrid Networks for Classification of Remotely-Sensed Satellite Images", in *Proceedings of the 1992 International Conference on Artificial Neural Networks (ICANN-92), Brighton, U.K.* (I. Aleksander and J. Taylor, eds.), vol. 2, pp. 1193–1196, Elsevier Science Publications (North-Holland), September 1992.

10. I. Kanellopoulos, A. Varfis, G. G. Wilkinson, and J. Mégier, "Neural network classification of multi-date satellite imagery", in *Proceedings of the International Geoscience and Remote Sensing Symposium (IGARSS'91), Espoo, Finland*, vol. IV, pp. 2215–2218, IEEE Press, June 1991.

11. I. Kanellopoulos, A. Varfis, G. G. Wilkinson, and J. Mégier, "Land cover discrimination in SPOT HRV imagery using an artificial neural network—a 20-class experiment", *International Journal of Remote Sensing*, vol. 13, no. 5, pp. 917–924, 1992.

12. I. Kanellopoulos, G. G. Wilkinson, and A. Chiuderi, "Land cover mapping using combined Landsat TM imagery and textural features from ERS-1 Synthetic Aperture Radar Imagery", in *Image and Signal Processing for Remote Sensing* (Jacky Desachy, ed.), Proc. SPIE 2315, pp. 332–341, September 1994.

13. J. Key, A. Maslanic, and A. J. Schweiger, "Classification of merged AVHRR and SMMR arctic data with neural network", *Photogrammetric Engineering and Remote Sensing*, vol. 55, no. 9, pp. 1331–1338, 1989.

14. R. Kwok, Y. Hara, R. G. Atkins, S. H. Yueh, R. T. Shin, and J. A. Kong, "Application of neural networks to sea ice classification using polarimetric SAR images", in *Proceedings of the International Geoscience and Remote Sensing Symposium (IGARSS'91), Espoo, Finland*, vol. I, pp. 85–88, IEEE Press, June 1991.

15. Z. K. Liu and G. G. Wilkinson, "A neural network approach to geometrical rectification of remotely-sensed satellite imagery", Technical Note 1.92.118, Institute for Remote Sensing Applications, Joint Research Centre, Commission of the European Communities, Ispra, Italy, 1990.

16. G. Loung and Z. Tan, "Stereo matching using artificial neural networks", *International Archives of Photogrammetry and Remote Sensing*, vol. 29, no. B3, pp. 417–421, 1992.

17. A. Maslanic, J. Key, and A. J. Schweiger, "Neural network identification of sea-ice seasons in passive microwave data", in *Proceedings of the International Geoscience and Remote Sensing Symposium, (IGARSS'90), Maryland, USA*, pp. 1281–1284, IEEE press, Piscataway, NJ, 1990.

18. L. E. Pierce, K. Sarabandi, and F. T Ulaby, "Application of an artificial neural network in canopy scattering inversion", *International Journal of Remote Sensing*, vol. 15, no. 16, pp. 3263–3270, 1994.

19. D. E. Rumelhart, G. E. Hinton, and R. J. Williams, "Learning internal representations by error propagation", in *Parallel Distributed Processing. Explorations in the Microstructures of Cognition, Vol. 1: Foundations* (D. E. Rumelhart, J. L. McClelland, and the PDP Research Group, eds.), pp. 318–362, Cambridge, Massachusetts: MIT Press, 1988.

20. T. W. Ryan, P. J. Sementelli, P. Yuen, and B. R. Hunt, "Extraction of shoreline features by neural nets and image processing", *Photogrammetric Engineering and Remote Sensing*, vol. 57, no. 7, pp. 947–955, 1991.

21. M. S. Spina, M. J. Schwartz, D. H. Staelin, and A. J. Gasiewski, "Application of multilayer feed-forward neural networks to precipitation cell-top altitude estimation", in *Proceedings of the International Geoscience and Remote Sensing Symposium (IGARSS'94), Pasadena, California*, vol. IV, pp. 1870–1872, IEEE Press, Piscataway, NJ, August 1994.

22. G. G. Wilkinson, F. Fierens, and I. Kanellopoulos, "Integration of neural and statistical approaches in spatial data classification", *Geographical Systems*, vol. 2, pp. 1–20, 1995.
23. G. G. Wilkinson, S. Folving, I. Kanellopoulos, N. McCormick, K. Fullerton, and J. Mégier, "Forest Mapping from multi-source satellite data using neural network classifiers — An experiment in Portugal", *Remote Sensing Reviews*, vol. 12, pp. 83–106, 1995.
24. G. G. Wilkinson, I. Kanellopoulos, W. Mehl, and J. Hill, "Land Cover Mapping Using Combined Landsat Thematic Mapper Imagery and ERS-1 Synthetic Aperture Radar Imagery", in *Proceedings 12th PECORA Symposium: "Land Information from Space Based Systems", Sioux Falls, South Dakota, 24–26 August 1993*, pp. 151–158, American Society for Photogrammetry and Remote Sensing, 1994.
25. M. Zhang and R. A. Scofield, "Artificial neural network technique for estimating heavy convivrainfall an recognising cloud mergers from satellite data", *International Journal of Remote Sensing*, vol. 15, no. 16, pp. 3241–3261, 1994.

A Comparison of the Characterisation of Agricultural Land Using Singular Value Decomposition and Neural Networks

S. Danaher[1], G. Herries[1], T. Selige[2], and M. Mac Súirtán[3]

[1] Leeds Metropolitan University Leeds LS1 2ET
[2] GSF - Centre for Environmental and Health Research
   Section PUC, PO. BOX 1129,D-85758, Oberschieissheim, Germany
[3] University College Belfield Dublin 4 Ireland

Summary. Methods are defined and tested for the characterisation of agricultural land from multi-spectral imagery, based on Singular Value Decomposition (SVD) and Artificial neural networks (ANN). The SVD technique, which bears a close resemblance to multivariate statistic techniques, has previously been successfully applied to problems of signal extraction for marine data [1] and forestry species classification [2].

In this study the two techniques are used as a classifier for agricultural regions, using airborne Daedalus ATM data, with lm resolution. The specific region chosen is an experimental research farm in Bavaria, Germany. This farm has a large number of crops, within a very small region and hence is not amenable to existing techniques. There are a number of other significant factors which render existing techniques such as the maximum likelihood algorithm less suitable for this area. These include a very dynamic terrain and tessellated pattern soil differences, which together cause large variations in the growth characteristics of the crops.

Both the SVD and ANN techniques are applied to this data set using a multi-stage classification approach. Typical classification accuracy's for the techniques are of the order of 85-100%. Preliminary results indicate that the methods provide fast and efficient classifiers with the ability to differentiate between crop types such as Wheat, Rye, Potatoes and Clover.

1. Overview of the classification problem

Assuming the image consists of pixels of a number of different pure classes. It is convenient to divide the classes into two types: class a and class b. Class a will consist of all the pixels we are interested in such as a particular crop type. Class b will consist of all the other pixels in the image, which may consist of many data types. For classification we wish to find an optimal technique which will separate the two populations. One method for separating the two classes is to have some sort of transform which will map all the pixels of class a onto say 1 and all the pixels of class b onto say 0. Ideally all the pixels of class a would have a score of 1 and the background pixels another score of 0. Unfortunately it is unlikely that any algorithm will produce such a perfect separation. In practice the score values for both the pixel types will be spread around the two means. These score values are often plotted as histograms and the technique known as histogram separation (c.f. figure 1.1).

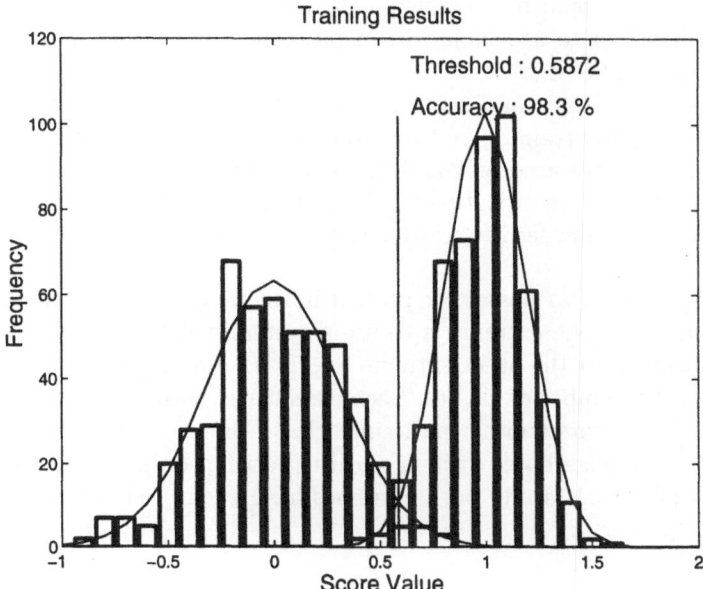

Fig. 1.1. A typical Histogram Separation

The goodness of an estimator may be defined by how closely it maps the two pixel types onto the respective values of 0 and 1. As the estimator improves, the width of the standard deviations, (or variances), of the two histograms will decrease. It is natural therefore to define the best estimator as the one which minimises the standard deviations of the two score values as much as possible. The optimal estimator is the one which will map the respective pixel types onto 1 and 0, subject to the minimisation of

$$e = \frac{1}{n_a} \sum_{i=1}^{i=n_a} (s_{a_i} - 1)^2 + \frac{1}{n_b} \sum_{i=1}^{i=n_b} (s_{b_i})^2 \tag{1.1}$$

where n_a is the number of pixels of the class in which we are interested and n_b is the number of background pixels and s_a and s_b are the scores corresponding to particular images. This is often referred to as minimisation in the least squares sense. The search for a good classifier is one which produces minimum error. A threshold value can now be applied, typically with a value of about 0.5. The pixels with a score above this value are considered to be signal and those below background. More formally, one possible approach would be to assume that the score values are normally distributed for both the signal and background data it can be shown that optimal value at which to locate the threshold (that which will give the greatest classification accuracy) is given by

$$\frac{n_a}{\sigma_a^2} exp \left(\frac{(s - \mu_a)^2}{2\sigma_a^2} \right) = \frac{n_b}{\sigma_b^2} exp \left(\frac{(s - \mu_b)^2}{2\sigma_b^2} \right) \tag{1.2}$$

Other methods would be to define a threshold based on the distribution of the data itself, or to use some sort of probabilistic measure based again on the statistics of the data.

Furthermore it is standard practice to divide the data set into two; one used for algorithm training and the other verification. A number of different methodologies were used on the data: Singular Value Decomposition (SVD) a Learning Vector Quantisation (LVQ) artificial neural network architecture and the very popular feed forward network artificial neural network architecture.

As more than two classes are present in the image a number of strategies may be used. At one extreme is to use a system which classifies every class against at once. At the other extreme $(c-1)^2/2$ different classifiers are used where c is the number of classes. Each classifier is optimised to separate one particular class from another particular class. The classifiers may then form part of a knowledge based system and the decision based on the $(c-1)^2/2$ score values. In this study both extremes have been used depending on the methodology.

2. Singular value decomposition

Singular value decomposition (SVD) represents a powerful numerical technique for the analysis of multivariate data [3,4]. SVD can be used as a preliminary stage in most types of multivariate analysis, and can greatly increase the computational efficiency of linear techniques such as key vector analysis, and non linear techniques such as cluster analysis and neural network analysis. SVD is also an extremely effective technique for the reduction of white noise. The inherent attributes of the SVD technique may have a considerable influence on the dataset, the more important of which may be summarised as follows.

i. Dimensional Reduction. Unless the parameters (bands) are completely independent or the data set is totally dominated by noise, SVD will determine a linear transformation which will convert the parameters into totally independent variables. Furthermore, it will do so in such a way that the SVD parameters (which are linear combinations of the original parameters) are chosen in decreasing order of significance. Transforming to SVD parameters is particularly important in neural network type analysis as the number of connections (determined by the number of layers) grows dramatically with the number of parameters.

ii. Invertibility. A data set which has undergone SVD can be regenerated by simple matrix multiplication.

iii. Data Compression. The data set can be regenerated with high accuracy, by ignoring parameters (dimensions) which make minimal contribution.

iv. Noise Reduction. No data set is totally noise free. If the parameterised data is considered to form an n dimensional space, where n is the number of parameters, white noise will spread uniformly over the space, having no preferred direction. An orthogonal transformation, such as SVD, is equivalent to a rotation/reflection in n-space. Dimensions which are deliberately excluded will have exactly the same noise content as those which are included (within statistical fluctuations). The signal to noise level is improved by discarding the higher dimensions. Using SVD for data compression reduces the noise level of the data.

v. Orthogonality. If the data is subsequently used for a cluster type analysis the fact that the SVD parameters are totally independent allows separation of events to be defined (if the 2-norm is used) as

$$r = \sqrt{\sum_{i=1}^{i=k} (w_i - < w_i >)^2} \tag{2.1}$$

where r is the normalised distance from the centroid of a cluster to every point, k is the number of included SVD dimensions and w_i are the individual SVD parameters. The standard cluster analysis technique uses the equivalent 2-norm definition for a non orthogonal basis set

$$r = \sqrt{\sum_{i=1}^{i=n} \sum_{j=1}^{j=n} \mathbf{A}_{ij}(b_i - < b_i >)(b_j - < b_j >)} \tag{2.2}$$

where r is the normalised distance from the centroid of a cluster to every point, b_i are the individual band parameters and $\mathbf{A}$ is the inverse of the variance-covariance matrix.

vi. Efficiency. SVD is an extremely efficient and robust technique. For a data set of 10000 signal pixels and 10000 background pixels and using seven bands, a SVD takes less than one minute on a 486 PC.

vii. Key Vector analysis. SVD allows a technique called key vector analysis to be used. Whereas key vector analysis is not expected to be as effective as other more sophisticated techniques such as cluster analysis or neural network analysis it has a number of advantages:

1) The key vector is unique.

2) The key vector can be determined rapidly using a set of Monte-Carlo simulations or a subset of the data.

3) Once the key vector has been obtained a score value can be determined on the original data set with only n floating point operations per event, where n is the number of bands used. It is therefore far more computationally efficient than most of the other techniques used.

4) Due to the fact that the key vector is highly constrained, the likelihood of spurious results is significantly reduced.

Before proceeding to a formal definition of SVD it is worthwhile stating that it bears a very close relationship to Characteristic Vector analysis [5]. The difference is in the elegance of the formalism. If $\mathbf{O}$ is a $m \times n$ data matrix the SVD of $\mathbf{O}$ is defined by:

$$\mathbf{O} = \mathbf{WLV} \tag{2.3}$$

The matrices $\mathbf{W}$, $\mathbf{L}$ and $\mathbf{V}$ can be defined as follows: $\mathbf{V}$ is a $n \times n$ matrix containing the unit eigenvectors of $\mathbf{O}^T\mathbf{O}$, (the variance/covariance matrix), on each row. These are sorted in descending significance. If the matrix $\mathbf{O}$ has had its mean subtracted (column by column) the matrix $\mathbf{V}$ will be a matrix of the characteristic vectors as defined in multivariate statistics. $\mathbf{L}$ is a $m \times n$ matrix of the form:

$$\mathbf{L} = \begin{pmatrix} D & 0 \\ 0 & 0 \end{pmatrix} \tag{2.4}$$

where $\mathbf{D}$ is diagonal. The diagonal elements of $\mathbf{D}$ are the square roots of the eigenvalues of $\mathbf{O}^T\mathbf{O}$ sorted in descending order.

The matrix $\mathbf{W}$ is a $m \times m$ containing the unit eigenvectors of $\mathbf{OO}^T$ on each column. It also may be considered as a matrix containing the weights or scalar multipliers of the characteristic vectors $\mathbf{V}$ in the data matrix $\mathbf{O}$.

There is also an 'economy sized' SVD where $\mathbf{W}$ is $m \times n$, $\mathbf{L}$ is $n \times n$ and $\mathbf{V}$ is $n \times n$. For remote sensing data, m (the number of pixels) can be very large and hence a $m \times m$ matrix would be too large to be manageable. Whereas all the eigenvectors will generally be needed to define $\mathbf{O}$ exactly, in many cases a sufficiently good approximation to $\mathbf{O}$ can be achieved by taking only a few eigenvectors. Then:

$$\mathbf{O} \approx \mathbf{WLV} \tag{2.5}$$

with $\mathbf{W}$ being $m \times k$, $\mathbf{L}$ being $k \times k$ and $\mathbf{V}$ being $k \times n$; taking the first k columns of $\mathbf{W}$, the first k rows and columns of $\mathbf{L}$ and the first k rows of $\mathbf{V}$.

Generation of the Key Vector

The first step in a two stage process involves finding that key vector which is most efficient at separating a mixed signal population and background population. (The background will consist of all the non signal regions in the image) Due to the high variability of background classes, the technique may have to be applied a number of times; filtering the pixels classified as signal. The second stage of the operation is to consider the desired species as one population and the other types of ground cover as a background class. Each pixel is considered to be a vector in n dimensional space where n is the number of bands.

$$\mathbf{o} = (h_1, h_2, h_3, .., h_n) \tag{2.6}$$

The Data matrix $\mathbf{O}$ was generated by combining a set of signal and background classes.

$$\mathbf{O} = \begin{pmatrix} h_{11} & h_{12} & h_{13} & ... & h_{1n} \\ h_{21} & h_{22} & h_{23} & ... & h_{2n} \\ . & . & . & . \\ h_{m1} & h_{m2} & h_{m3} & ... & h_{mn} \end{pmatrix} \tag{2.7}$$

In SVD we wish to find an optimal single parameter (which will be a linear combination of the original pixel parameters) which will separate the two populations. This optimal choice of parameter weights $(a_1, a_2 ... a_n)$, can be considered as a vector in n dimensional space. This direction defines the key vector [6]. It is the direction in multidimensional space which provides optimal separation of the two populations.

The key vector is obtained using singular value decomposition. We note that the definition given is equivalent to finding a vector $\mathbf{k}$ such that

$$\mathbf{Ok} = \mathbf{a} \tag{2.8}$$

where $\mathbf{a}$ is a column vector of size $1 \times m$. The nth element of $\mathbf{a}$ is one if the corresponding row in $\mathbf{O}$ contains signal parameters and zero otherwise. That is, the scalar product of the multi-dimensional vector associated with each event and the key vector, will be 1 if the pixel contains forest and 0 if the pixel contains background. Therefore

$$\mathbf{WLVk} = \mathbf{a} \tag{2.9}$$

or

$$\mathbf{k} = \mathbf{V}^{-1}\mathbf{L}^{-1}\mathbf{W}^{-1}\mathbf{a} \tag{2.10}$$

Note as both $\mathbf{V}$ and $\mathbf{W}$ are orthogonal these matrices can be inverting by taking their transpose. Also as $\mathbf{L}$ is diagonal, $\mathbf{L}^{-1}$ can be obtained by taking the reciprocal of the diagonal terms. Therefore once the SVD has been calculated there is little computational overhead in obtaining the key vector.

In practice it is usual to constrain the number of included dimensions to those in which the signal to noise ratio is reasonable. This will reduce the accuracy of the solution, but render it more robust with respect to noise and spurious variations in a small number of pixels. The optimal key vector $\mathbf{k}$ is now applied to the data matrix $\mathbf{O}$ in order to compute a vector $\mathbf{s}$ of m score values, one for each pixel.

$$\mathbf{s} = \mathbf{Ok} \tag{2.11}$$

This score value may now be used for classification using the techniques described in section 1.

3. Artificial neural networks (ANN)

Neural networks are becoming ever popular as classifiers for remote sensing data [7,8]. One of their main advantages, is they make no assumption about the statistical distribution of the data, whereas techniques such as the Maximum likelihood algorithm assume the data has a known statistical distribution. One of the difficulties with ANN is deciding on a suitable training data set, which is correlated with the features you wish to extract.

We have used the BP algorithm to train the ANN, as this particular algorithm is well suited to feature/parameter estimation. Two BP algorithms will be presented, the standard BP algorithm based on Gradient descent and an approximation of Newton's method called Levenberg-Marquardt.

Back-propagation - Gradient Descent

The BP learning rule is used to adjust the weights and biases of networks in order to minimise the sum-squared error of the network. This is done by continually changing the values of the network weights and biases in the direction of steepest descent with respect to error, hence the name Gradient Descent. Derivatives of error (called Delta vectors) are calculated for the networks output layer and then back-propagated through the network until the error derivatives are available for each layer. The weights/biases of a network are adaptively changed as described above, but if the data set describes a complex function (as remote sensing data often does), then the error surface for the network may also be very complex with a number of local minima which bear little relationship with the global minima (which is the target for the training procedure). Simple BP is very slow because it requires small learning rates for stable learning. Back-propagation is not guaranteed to find an optimal solution and is often prone to getting trapped in local minima.

Momentum. An adaptation of the simple BP technique is to use a momentum feature which allows the network to respond not only to the local gradient, but also to recent trends in the error surface. This has the effect of acting like a low pass filter, which allows the network to ignore small features in the error surface, so that the network is less prone to becoming trapped in local minimum.

Adaptive Learning. A further adaptation to reduce the training time is to implement an adaptive learning rate, which monitors the errors between iterations (as with momentum), and if the new error exceeds the old error by more than a predefined ratio, the new weights and biases are discarded and the learning rate is decreased. If the new error is less than the old error, the learning rate is increased, to the extent that the network can learn without large error increases. Thus a near optimal learning rate is obtained for the local error surface.

This adaptation will be called BP with Momentum and Adaptive learning (BPMAL) throughout this paper. These two additions significantly increase the efficiency of BP, but these techniques do not guarantee that the network will not become trapped in local minima. A more advanced technique which is efficient at finding a solution is the Levenberg-Marquardt approximation technique.

Back-propagation - Levenberg Marquardt

Back-propagation using Levenberg-Marquardt (BPLM) is an improvement on the gradient descent technique used by BPMAL and uses an approximation of Newtons method called Levenberg-Marquardt. This technique is more powerful than gradient descent and converges on the global minimum much faster, but is very memory intensive.

The Levenberg-Marquardt update rule is:

$$\Delta \mathbf{W} = (\mathbf{J}^T \mathbf{J} + \mu \mathbf{I})^{-1} \mathbf{J}^T \mathbf{e} \tag{3.1}$$

where $\mathbf{J}$ is the Jacobian matrix of derivatives of each error to each weight, μ is a scalar and $\mathbf{e}$ is an error vector. If the scalar μ is very large, equation 3.1 approximates gradient descent, however, if it is small the expression becomes the Levenberg-Marquardt approximation. This method is faster than BPMAL, but is less accurate when near an error minima. The scalar μ is adjusted in a similar manner to the adaptive learning rate in BPMAL. Providing the error gets smaller, μ is made bigger and conversely if the error increases then μ is made smaller.

Both BPMAL and BPLM are sensitive to the number of neurons in each layer and also the number of hidden layers in the network. While generally the more neurons in hidden layers the more accurately the network can fit the data, too many neurons/hidden layers can cause overfitting to occur and the network will lose its generalisation capabilities.

Learning Vector Quantisation (LVQ) Neural Networks

Learning Vector Quantisation [9] is way of training competitive layers in a supervised manner. The network consists of three layers: a layer of Instar Neurons, a competitive layer and a linear layer (figure 3.1). In essence the Instar Neurons compare an input pattern with a stored pattern and gives an output which is proportional to the pattern match. The competitive layer chooses the Instar Neuron with the highest response. The linear layer maps the output of the competitive layer onto the desired classes.

An Instar Neuron has a transfer function

$$y = \sum_{i=1}^{i=n} w_i x_i = |\mathbf{w}||\mathbf{x}|cos(\theta) \tag{3.2}$$

where y is the output, w the stored weights n the number of dimensions and x the input. The output therefore is equivalent to taking a scalar product between the vector input and the stored weights. if **w** and **x** are normalised to length 1 the output will depend on the cosine of the angle and therefore will be a maximum of one if the two vectors are pointing in the same direction in n dimensional space. These neurons are trained using the Kohonen learning rule:

$$\Delta\mathbf{w} = k(\mathbf{x} - \mathbf{w}) \tag{3.3}$$

where k is the learning rate, using only the inputs for which the output y has the highest response. The Instar neurons will organise themselves into natural classes with one neuron being assigned to each class. At this stage the network is unsupervised. After training the linear layer is utilised to map the natural classes to the desired classes.

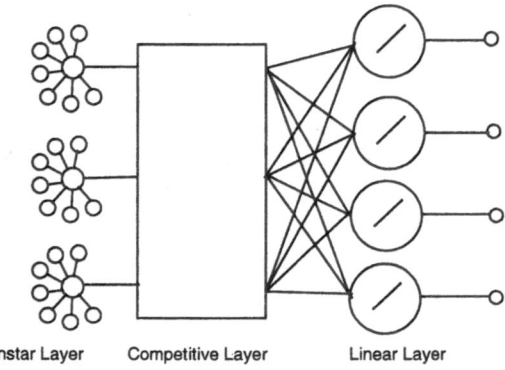

Instar Layer   Competitive Layer   Linear Layer

Fig. 3.1. The LVQ Network

4. Classification strategy on the ATM data

The ATM Data

The Daedalus airborne scanner was mounted in a DO228 aircraft and flown over the experimental farm at an altitude of 450 meters. This produced an effective pixel resolution of 1m. The image has been panoramically corrected. The image was acquired on the 4th July 1994 at 12.00am.

Scheyern experimental farm is managed under the project Forschungsverbund Agrarökosysteme München, which is a co-operation between GSF - Forschungszentrum für Umwelt und Gesundheit and several institutes of the Technische Universität München. The site is situated about 40 kilometres north of Munich in a hilly landscape derived from tertiary sediments. The area amounts to 143 ha. (10% pasture, 90% cropland). It illustrates the typical problems of an intensively farmed landscape: Erosion, soil compaction and contamination of ground water. The area covered by the experimental farm has a very dynamic terrain and is situated between 450 and 490 meters above sea level. The land cover types are representative of much of Western Europe. Cultivation is more difficult on the slopes and erosion is much higher. The annual precipitation is approximately 833 millimetres.

Classification Strategy

For the SVD and ANN techniques, representative homogeneous pixel training sets were extracted from airborne Daedalus ATM data, with lm resolution. A verification data set was also extracted to verify that over-training had not occurred and the networks could still generalise. Sample sizes for the training and verification sets were 520 and 720 multispectral pixels comprising 12 bands (see table 4.1), for each class respectively. The training goal for SVD and the ANN's is to map the data set to an optimum score value for the desired and undesired classes, which is 1 and 0 respectively. It is important to note that whereas every effort was made to ensure that the training and verification data sets were identical from the point of view of ground cover, they are from different sections of the image and are not interlaced.

Table 4.1. Daedalus Channels

Channel Number	Wavelength (mm)
1	0.420 - 0.450
2	0.450 - 0.520
3	0.520 - 0.600
4	0.605 - 0.625
5	0.630 - 0.690
6	0.695 - 0.750
7	0.760 - 0.900
8	0.910 - 1.050
9	1.550 - 1.750
10	2.080 - 2.350
11	8.500 - 13.000
12	8.500 - 13.000

SVD. Individual key vectors are produced for every class against every other class using the first three characteristic vectors. This is an excessive method, but it does ensure that the vector space is parameterised fully. These key vectors are then used to filter the data set to characterise the various land cover classes. This is a modular approach and allows extra land cover classes to be characterised, without having to regenerate all the key vectors previously produced.

Feed Forward Neural Networks. Again a modular approach is proposed with a neural network for each class. Again to fully parameters the vector space, such that random fluctuations will not cause misclassifications a ANN must be generated for every class against every other class. If an extra land cover class is subsequently added then ANNs for that particular class against every other class need be trained. The networks were trained on dimensionally reduced data using the first four characteristic vectors.

LVQ Networks. The LVQ architecture is by nature a natural vector classifier. A single network was used for classification and the number of Instar neurons varied to give optimal results. The networks were trained on dimensionally reduced data using the first four characteristic vectors.

5. Results

Both of the artificial neural network strategies were optimised by varying in the case of the LVQ network the number of Instar Neurons and in the case of the Feed Forward network, the number of neurons in the hidden layer. In the case of the LVQ network, optimal results on the verification data were produced using 11 neurons. In the case of the Feed Forward Networks optimal results were produced using 5 neurons in the hidden layer.

The SVD proved to be the most accurate of the three techniques with an accuracy of 94% (table 5.1) followed by the LVQ network with an accuracy of 90% (table 5.2) and the Feed Forward architecture with an accuracy of 77% (table 5.3). Whereas the Feed forward architecture gave perfect results on the training data it was unable to generalise. This may be to some extent due to the fact that whereas the training and verification data sets come from the same image, they are from separate areas may therefore be somewhat different in their statistics. All three techniques have an accuracy of over 97% on the training data.

It is interesting to note that whereas the SVD and LVQ techniques have comparable accuracy there is a 14% difference between the classification results for the verification data between the two techniques. This would indicate that further improvements might be made by using a combination of the two techniques. Such as highlighting pixels in which the classification differed and using another classification technique. Unfortunately the accuracy was so high with the training data that the number of differently classified pixels was less than 1% and hence was too small a training set as to be useful.

6. Conclusions

As can be seen in section 5, the results are extremely promising for the classification of this type of remote sensing data especially for the LVQ networks

Table 5.1. Results of SVD/Key Vector Analysis Classification

	Ground Class 1	Ground Class 2	Ground Class 3	Ground Class 4	Ground Class 5	Ground Class 6	Ground Class 7
Output Class 1 - Clover	644	0	0	0	11	0	0
Output Class 2 - Lupine	2	582	0	46	2	0	0
Output Class 3 - Potato	0	7	699	8	0	0	7
Output Class 4 - W.Wheat	0	125	1	666	0	0	0
Output Class 5 - SunFlower	74	6	0	0	707	0	0
Output Class 6 - Rye	0	0	0	0	0	720	0
Output Class 7 - Grassland	0	0	20	0	0	0	711
Output Class Unclassified	0	0	0	0	0	0	2
Total Pixels	720	720	720	720	720	720	720

Table 5.2. Results of the LVQ Classification

	Ground Class 1	Ground Class 2	Ground Class 3	Ground Class 4	Ground Class 5	Ground Class 6	Ground Class 7
Output Class 1 - Clover	535	0	0	0	1	0	0
Output Class 2 - Lupine	0	680	0	85	1	0	0
Output Class 3 - Potato	0	21	564	8	0	0	7
Output Class 4 - W.Wheat	0	13	1	627	0	0	0
Output Class 5 - SunFlower	185	6	0	0	718	0	0
Output Class 6 - Rye	0	0	0	0	0	720	0
Output Class 7 - Grassland	0	0	150	0	0	0	718
Output Class Unclassified	0	0	0	0	0	0	0
Total Pixels	720	720	720	720	720	720	720

Table 5.3. Results of the Feed Forward Network Classification

	Ground Class 1	Ground Class 2	Ground Class 3	Ground Class 4	Ground Class 5	Ground Class 6	Ground Class 7
Output Class 1 - Clover	624	0	0	0	0	0	0
Output Class 2 - Lupine	0	708	1	456	7	0	0
Output Class 3 - Potato	0	10	193	0	0	1	2
Output Class 4 - W.Wheat	0	1	0	228	0	0	0
Output Class 5 - SunFlower	96	1	0	0	713	0	0
Output Class 6 - Rye	0	0	0	0	0	719	0
Output Class 7 - Grassland	0	0	526	13	0	0	718
Output Class Unclassified	0	0	0	23	0	0	2
Total Pixels	720	720	720	720	720	720	720

and SVD/Key vector analysis. Although these results are for a best case scenario of selected training and verification areas, the key vectors are still quite robust when applied to whole images. There are as ever some problematic areas, which are mainly due to the resolution of the scanner and localised soil differences, however the mis-classifications that occur are due to insufficient representation of sub-classes. In areas such as the one chosen for this study, where their are a large number of different crops with very different growth characteristics, a large number of key vectors for each crop type would have to be generated. This is the main downfall of the SVD technique when there are a large number of classes to be classified. Therefore future work will concentrate on integrating other techniques with SVD, to allow a first pass of the imagery with a classifier such as an LVQ Neural network.

Another problem area is the use of hard thresholding in the classification. This is far from being the best decision criteria. Current work is investigating other possible decision criteria, including the use of fuzzy logic. Fuzzy logic would allow a tolerance band for the score values, which is a more sensible route for operational classification.

Future work will concentrate on building a knowledge base of crop key vectors, crop sub-classes and developing an intelligent classification system that can classify mixed pixels and crop characteristics, such as Yield and Biomass. The robustness of the key vectors will also be tested on imagery with different acquisition dates and varying resolutions.

References

1. O. Zalloum, E. O'Mongain, J. Walsh, S. Danaher, and L. Stapleton, "Dye concentration estimation by remotely-sensed spectral radiometry", *International Journal of Remote Sensing*, vol. 14, pp. 2285–2300, 1993.
2. S. Danaher, G. Herries, M. MacSiurtain and E. O'Mongain, "Classification of forestry species using singular value decomposition", SPIE, 2314–56, pp. 270–280, 1994.
3. W. H. Press, B. P. Flannery, S. A. Teukolsky, and W. T. Vettering, *Numerical Recipes - The Art of Scientific Computing*, Cambridge Univ. Press, Cambridge, 1986.
4. G. H. Golub, and C. F. Van Loan, *Matrix Computations*, Johns Hopkins Univ. Press, Baltimore, MD, 1983.
5. J. L. Simonds, "Applications of characteristic vector analysis to photographic and optical response data", *J.Optical.Soc.America*, vol. 53, pp. 968–975, 1963.
6. S. Danaher and E. O'Mongain, "Singular value decomposition in multispectral radiometry", *International Journal of Remote Sensing*, vol. 13, pp. 1771–1777, 1992.
7. J. A. Benediktsson, P. H. Swain, and O. K. Ersoy, "Neural network approaches versus statistical methods in classification of multisource remote sensing data", *IEEE Transactions on Geoscience and Remote Sensing*, vol. 28, no. 4, pp. 540–552, 1990.

8. J. A. Benediktsson, P. H. Swain, and O. K. Ersoy, "Conjugate-gradient neural networks in classification of multisource and very-high-dimensional remote sensing data", *International Journal of Remote Sensing*, vol. 14, no. 15, pp. 2883–2903, 1993.
9. T. Kohonen, *Self Organisation and Associative Memory*, 2nd edition, Berlin: Springer-Verlag, 1987.

Land Cover Mapping from Remotely Sensed Data with a Neural Network: Accommodating Fuzziness

Giles M. Foody

Telford Institute of Environmental Systems,
Department of Geography, University of Salford,
Salford, M5 4WT, United Kingdom.

Summary. Neural networks are attractive for the supervised classification of remotely sensed data. There are, however, many problems with their use, restricting the realisation of their full potential. This article focuses on the accommodation of fuzziness in the classification procedure. This is required if the classes to be mapped are continuous or if there is a large proportion of mixed pixels. A continuum of fuzzy classifications was proposed and it is shown that a neural network may be configured at any point along this continuum, from a completely-hard to a fully-fuzzy classification. Examples of fuzzy classifications are given illustrating the potential for mapping continuous classes and reducing the mixed pixel problem.

1. Introduction

Remotely sensed data may be used in three major application areas in the study of the terrestrial environment. Arranged along a generally perceived scale of increasing difficulty, these are the mapping, monitoring and estimation of environmental phenomena. In each of these application areas neural networks may be used in the extraction of the desired information from the remotely sensed data. Here attention is focused on land cover mapping from remotely sensed data with a neural network.

Land cover is a fundamental variable that links aspects of the human and physical environments. While its importance is recognised, data on land cover are often out-of-date, of poor quality or inappropriate for a particular application [1]. Furthermore, land cover data are not easy to acquire [2]. This is particularly the case if data are required for large areas or if frequent up-dating is required. Often the only feasible approach to map land cover is through the use of remotely sensed data, especially for mapping at regional to global scales. Thus land cover mapping has become one of the most common applications of remote sensing and is often a perquisite to accurate monitoring and parameter estimation. Although often perceived as a relatively trivial application, land cover mapping from remotely sensed data is fraught with difficulty and has not yet reached operational status [3]. A number of reasons may be cited for the failure to realise the full potential of remote sensing as a source of land cover data. One set of factors relate to the methods used. Neural networks have considerable potential for resolving some of the problems and so facilitate accurate land cover mapping from remotely sensed data and this article discusses some important issues linked to realising their full potential in this application area.

2. Land Cover Mapping from Remotely Sensed Data

Typically a supervised classification is used in the mapping of land cover from remotely sensed data. This type of classification is generally applied on a per-pixel basis and has three distinct stages. First, the training stage, in which pixels of known class membership in the remotely sensed data are characterised and class 'signatures' derived. In the second stage, these training statistics are used to allocate pixels of unknown class membership to a class in accordance to some decision rule. Third, the quality of the classification is evaluated. This is generally based on the accuracy of the classification which is assessed by comparing the actual and predicted class of membership for a set of pixels not used in training the classification. The accuracy of maps derived from a supervised image classification is therefore a function of factors related to each of the training, allocation and testing stages of the classification. Although there has been considerable research on supervised classification, surprisingly little has questioned the validity of conventional classification techniques for mapping. This article focus on one major problem with conventional classification techniques and indicates how this may be resolved with a neural network.

3. Supervised Classification

Of the many classification techniques available the most widely used are conventional statistical algorithms such as discriminant analysis and the maximum likelihood classification. These aim to allocate each pixel in the image to the class with which it has the highest probability of membership. Problems with this type of classification, particularly in relation to distribution assumptions and the integration of ancillary data, especially if incomplete, acquired at a low level of measurement precision or possessing a directional component, prompted the development and adoption of alternative classification approaches. Thus, for instance, attention has turned recently to approaches such as those based on evidential reasoning [4] and neural networks [5, 6]. A feed-forward neural network is particularly suited to supervised classification as it may learn by example and generalise. Moreover, it may accommodate data at various levels of measurement precision and directionality directly and make no assumptions about the nature of the data. This may be particularly valuable when multi-source data form the basis of the classification.

3.1 Classification by an Artificial Neural Network

Of the range of network types and architectures [7] classification has generally been achieved with a basic layered, feed-forward network architecture, and only this type of network is considered here. Such networks may be envisaged as comprising a set of simple processing units arranged in layers, with each unit in a layer connected by a weighted channel to every unit in the adjacent layer(s). The number of units and layers in the artificial neural network are determined by factors relating, in part, to the nature of the remotely sensed data and desired classification, with an input unit for every discriminating variable and an output unit associated with each class in the classification.

Each unit in the network consists of a number of input channels, an activation function and an output channel which may be connected to other units in the network. Signals impinging on a unit's inputs are multiplied by the weight of the inter-connecting channel and are summed to derive the net input to the unit. This net input is then transformed by the activation function to produce an output for the unit. There are a range of activation functions that may be used but typically a sigmoid activation function is used [7].

The values for the weighted channels between units are not set by the analyst for the task at hand but rather determined by the network itself during training. The latter involves the network attempting to learn the correct output for the training data. A learning algorithm such as back-propagation is used which iteratively minimises an error function over the network outputs and a set of target outputs, taken from a training data set. Training begins with the entry of the training data to the network, in which the weights connecting network units were set randomly. These data flow forward through the network to the output units. Here the network error, the difference between the desired and actual network output, is computed. This error is then fed backward through the network towards the input layer with the weights connecting units changed in proportion to the error. The whole process is then repeated many times until the error rate is minimised or reaches an acceptable level, which may be a very time consuming process. Conventionally the overall output error is defined as half the overall sum of the squares of the output errors.

Once the error has declined to an acceptable level, which is often determined subjectively, training ceases and the network is ready for the classification of cases of unknown class membership. In the classification each case is allocated to the class associated with the output unit with the highest activation level. Typically the output from the network is a hard classification, with only the code (i.e., nominal value) of the predicted class of membership indicated for each pixel.

Numerous comparative studies in a diverse range of studies have assessed the performance of neural networks relative to other classification approaches. In general these comparisons reveal that neural networks may be used to classify data at least as accurately but typically more accurately than conventional statistical algorithms such as the maximum likelihood classification and alternative approaches such as evidential reasoning [e.g. 5]. The main drawback of neural networks reported is the computationally demanding and therefore slow training stage, although this may be reduced through the use of appropriate parallel hardware or the adoption of 'one-shot' training neural networks such as those based on radial basis functions. These and other issues are discussed widely in the literature and are not the focus of this article. Here the aim is to summarise some key findings of earlier work on the accommodation of fuzziness into the classification. This is a general problem, common to all classification based approaches to land cover mapping.

4. Accommodation of Fuzziness

Although there are many instances when conventional classification techniques have been used successfully for the accurate mapping of land cover, they are not always appropriate. One important limitation of conventional approaches to land cover mapping (including those based on neural networks) is that the output derived consists only of the code of the allocated class. This type of output is often referred to as being 'hard' or 'crisp' and is wasteful of information on the strength of class membership generated in the classification. This information on the strength of class membership may, for instance, be used to indicate the confidence that may be associated with an allocation on a per-pixel basis, indicating classification reliability [8, 9], or be used in post-classification processing and enable more appropriate and informed analysis by later users, particularly within a geographical information system. Perhaps a more important limitation of 'hard' classifications is that they were developed for the classification of classes that may be considered to be discrete and mutually exclusive, and assume each pixel to be pure, that is comprised of a single class. Unfortunately this is often not the situation in mapping land cover from remotely sensed data. The classes may, for instance, be continuous and intergrade gradually. This intergradation implies a spatial co-existence of classes that cannot be accommodated by a conventional 'hard' classification technique [10] with many areas of mixed class composition, particularly near imprecise or fuzzy class boundaries. Alternatively, even if the classes may be considered to be distributed over the landscape as a mosaic of discrete classes many pixels will represent an area comprised of two or more classes. This problem stems largely from the use of the pixel as the basic spatial unit. In terms of factors such as size, shape and location on the ground, the pixel is largely an arbitrary spatial unit. Often the area represented by a pixel crosses the boundaries of classes resulting in a pixel of mixed land cover composition. The mixed pixel problem will be particularly apparent with coarse spatial resolution data, although the precise proportion of mixed pixels is a function of the combined effect of the landscape mosaic and the sensor's spatial resolution. Irrespective of their origin, mixed pixels are a major problem in land cover mapping applications. For example, while a mixed pixel must contain at least two classes a 'hard' classification technique will force its allocation to one class. Moreover, depending on the nature of the mixture and its composite spectral response, the allocated class need not even be one of the pixel's component classes. Since mixed pixels may be abundant in remotely sensed data they may therefore be a major source of error in land cover classifications. Furthermore, the mixed pixel problem may be most significant in mapping land cover from coarse spatial resolution data sets that are the backbone of regional to global scale analyses, which are also those most dependent on remote sensing as a source of land cover data. The errors in the land cover map provided by a conventional classification approach may also propagate into other studies that use the map.

Conventional classification approaches cannot therefore accommodate mixed pixels and so may not provide a realistic or accurate representation of land cover. Recognition of the effect and significance of mixed pixels has therefore led to the derivation of fuzzy classifications which allow for partial and multiple class membership characteristic of mixed pixels [11]. This could be achieved by 'softening' the output of a 'hard' classification. For instance, measures of the strength of class membership, rather than just the code of the most likely class of membership, may be output. This type of output may be considered to be fuzzy, as an imprecise allocation may be made and a pixel can display membership to all classes. Although generally used to produce a hard classification the output of a neural network classification may be softened to provide measures of the strength of class membership [12]. While this may provide a more appropriate representation of land cover that may be considered to be fuzzy the fuzziness of the land cover being represented has often been overlooked. The use of a fuzzy classification therefore does not fully resolve the mixed pixel problem, it only provides a means of appropriately representing land cover that may be considered fuzzy at the scale of the pixel. Thus while the class allocation made by a fuzzy classification may appropriately accommodate mixed pixels it must be stressed that this is only one of the three stages in the classification process. Relatively little attention has, however, focused on the accommodation of mixed pixels in the training and testing stages of a supervised classification. In both of these stages of the classification the ground data on class membership are generally related to the remotely sensed data at the scale of the pixel and so may be fuzzy. Since a large proportion of image pixels may be mixed it is important that this be recognised and accommodated throughout the classification. Frequently, however, the only way the fuzziness of the land cover on the ground is accommodated is by deliberately avoiding mixed pixels in training and testing the classification. Thus although an image may be dominated by mixed pixels only pure pixels are selected for training. This typically involves selecting training sites from only very large homogeneous regions of each class in order to avoid contamination of training sites by other classes. Moreover, research on refining training sets has often focused on removing potentially mixed pixels from the training set. In testing the classification pure pixels only should be used as the conventional measures of accuracy assessment were designed for application with 'hard' classifications. As the majority of pixels may be mixed failure to include them in the accuracy assessment may result in an inappropriate and inaccurate estimation of classification accuracy. There is therefore a need to account for fuzziness in both the classification output and ground data in the training and testing stages of the classification.

In both the training and testing stages of a supervised classification mixed pixels are therefore considered undesirable but if land cover is to be mapped accurately and the map evaluated appropriately they may be unavoidable. The accommodation of mixed pixels in the whole classification process may, however, be relatively easily achieved with a neural network [12, 13].

4.1 Accommodating Fuzziness in the Training Stage

Mixed pixels may be used directly in the training of an artificial neural network. This is because in an artificial neural network the analyst has, unlike with a conventional classification, an ability to specify a mixed target output for training samples; with a conventional classifier each training sample is associated unambiguously with a single class. Thus provided the land cover composition of training pixels is known a neural network could be trained on mixed pixels and used to classify remotely sensed data [12]. Operationally the data on the class composition of the pixels may be derived from a finer spatial resolution image co-registered to the image to be classified. In the near future it may be possible to make use of data acquired from satellite sensors that will acquire data at two or more spatial resolutions simultaneously.

4.2 Accommodating Fuzziness in the Allocation Stage

By outputting solely the code of the class associated with the unit in the output layer with the highest activation level information on the magnitude of the activation level of the output units is wasted; in the same way that maximum likelihood classification is wasteful of information by discarding the probability of class membership [11]. Since the network produces units capable of interpolation, the activation level of network output units may be used to form a fuzzy classification. The activation level of an output unit indicates the strength of membership of a pixel to the class associated with the output unit. Typically the activation level of a unit lies on a scale from 0 to 1 which reflects the variation from extremely low to extremely high strength of membership to the class associated with the output unit. For each pixel the strengths of class membership derived in the classification may be related to its land cover composition. Ideally the measures of the strength of class membership would reflect the land cover composition of a pixel. Thus the output for a pure pixel, representing an area of homogeneous cover of one land cover class, should be a very high strength of membership to the actual class of membership and negligible strength of membership to other classes. Alternatively in the output for a mixed pixel the strengths of class membership derived should reflect the relative coverage of the land cover classes in the area represented by the pixel [14].

4.3 Accommodating Fuzziness in the Testing Stage

A statement of classification accuracy is an essential accompaniment to a land cover map derived from remotely sensed data. Many methods for assessing classification accuracy have been proposed but these techniques were generally designed for application to 'hard' classifications and their use in the evaluation of fuzzy classifications may be misleading and erroneous. There is therefore a need for measures of classification accuracy which go beyond the confusion matrix and can accommodate fuzziness [12]. A number of approaches have been suggested but these generally only accommodate fuzziness in either the classification or the ground data and do not allow the comparison of classifications. A simple alternative is to measure the distance between land cover on the ground the fuzzy land cover representation derived from the classification. These measures are simple to interpret, with a low distance corresponding to an accurate representation, and the significance of the difference between classifications may be assessed [13].

4.4 A Continuum of Classification Fuzziness

A continuum of classification fuzziness may be defined. At one end of this continuum are conventional 'hard' classifications in which each pixel is associated with a single class throughout. With this approach pure pixels are used in training and testing the classification in which each pixel is allocated unambiguously to a single class. At the other end of the continuum are fully-fuzzy classifications, those which accommodate fuzziness in all three stages. Between these extremes are classifications of variable fuzziness. It would, for instance, be possible to use pure pixels in training a fuzzy classifier and evaluate the accuracy using a conventional confusion matrix based measure. In this instance two stages of the classification are 'hard' and one 'fuzzy'. Furthermore, this example generally corresponds to the meaning of the term 'fuzzy classification' in the literature. Clearly it is not a fully fuzzy classification and often the data used in training and testing the classification, while assumed to be pure, are mixed. Various combinations of 'hard' and 'fuzzy' classification stages could be defined for a single classification and two examples are described to below. It would, however, be possible to configure a neural network classification at any point along this continuum from completely-hard to fully-fuzzy.

4.5 Fuzzy Classification Examples

Fuzzy classifications are particularly valuable when the classes under study are continuous. This is often the situation with natural and semi-natural vegetation classes that lie along environmental gradients and therefore gradually intergrade. While a 'hard' classification technique cannot adequately represent such classes, the output of a fuzzy classification should be able to model the transitions from class to class. This not only provides a more appropriate representation of the vegetation it also enables characterisation of the transitional area/boundary between classes. Therefore even if the final land cover map must depict boundaries between classes these could be labelled with data defining the boundary's properties [15] such as boundary sharpness and width. A neural network has considerable potential for modelling the transition from one class to another. Briefly, in an earlier study [10] a transect crossing the forest-savanna boundary in West Africa was defined on NOAA AVHRR data. A fuzzy classification of the pixels along this transect was then produced with a neural network using training statistics derived from pure pixels of the main land cover classes at the site; thus pure pixels were used in the training stage but the class allocation stage was fuzzy. The results showed at each end of the transect, the activation level of the output unit to the class associated with relevant end of the transect was generally very high and close to 1.0 and that to the other classes close to 0.0 with the total of the activation levels over all output units was correspondingly close to 1.0. Where the transect crossed a sharp class boundary, such as that between savanna and water, the activation level would typically switch from a very high value that was close to 1.0 to savanna and close to 0.0 for all others, to one that was close to 1.0 for water and 0.0 to all others. For the more gradual transition between forest and savanna two main features were identifiable. First, the activation level of the most activated output unit was markedly less than 1.0 and furthermore, in some instances, rather than a pixel displaying an almost exclusive membership to one class, the total strength of membership was increasingly partitioned between two or more classes. Second, the magnitude of the total strength of class membership deviated from 1.0 in the transitional area. Both features indicate the response of the network to pixels drawn from outside the core regions of the classes. As these pixels lie between the core regions of the forest and savanna classes they display less certain membership to each class. Thus, for example, moving along the transect from the savanna end, the strength of class membership to savanna declines as that to the forest class increases. Moreover, since the network was trained to recognise the class membership of pixels taken from the core areas of the classes its ability to recognise pixels outside of these areas declines. Associated with this the magnitude of the output unit activation level for a class declines away from its core area and the total magnitude of the output unit activation levels for pixels in the transitional zone, where there is least similarity to the class core areas, deviates from 1.0. The zone where these 'anomalous' activation levels derived from the neural network occurred defined the transitional area between the classes. Furthermore, the activation levels enabled important characteristics of the boundary, such as its width, contrast and distance of temporal migration to be measured [10].

As noted above, a fuzzy classification may be required for discrete classes if the combined effect of the sensor's spatial resolution and the fabric of the landscape is a large proportion of mixed pixels. In another study [14] a fully-fuzzy classification was used to map broad land cover classes (forest, pasture and water) from NOAA AVHRR data of Brazil. The data set was dominated by mixed pixels, only some 3-4% of the image extract was composed of pure pixels. Using the land cover composition of the AVHRR pixels, derived from a co- registered Landsat TM image, mixed pixels were used to train and test a fuzzy classification; all three stages of the classification procedure therefore were fuzzy. Briefly the results demonstrated that mixed pixels could be accommodated throughout the classification process and derive a product that closely modelled the land cover on the ground. The spatial distribution of the classes could be visualised through fraction images and these corresponded closely to the landscape pattern evident in the Landsat TM data. Furthermore, quantitative evaluations of the classification, on both a per-class and overall basis, revealed that it provided an accurate representation of the land cover. Of particular significance was that the estimated areal extent of the classes at the site were much closer to their actual extent than would have been derived from a 'hard' classification. Since the method essentially involved estimating the sub-pixel land cover composition of pixels with a neural network using training data comprised of mixed pixels it also corresponds essentially to a non-linear mixture model that does not require problematic end-member spectra [14].

5. Summary and Conclusions

Neural networks are very attractive for supervised classification. The conventional 'hard' classification approach may, however, be inappropriate for mapping land cover as this may often be characterised by a degree of fuzziness. It has been demonstrated that fuzziness can be accommodated in all three of the stages of a neural network classification. Moreover, a continuum of classification fuzziness may be characterised and a neural network configured for classification at any point along this continuum. This may help enable a fuller realisation of the potential of neural networks for land cover classification.

Acknowledgement. This article has drawn on earlier research that benefited from the input of a range of people and their assistance is gratefully acknowledged. In particular I am grateful to Doreen Boyd for helpful discussions and the organisers of the COMPARES workshop for the invitation to present this paper.

References

1. Townshend, J., Justice, C., Li, W., Gurney, C., and McManus, J., "Global land cover classification by remote sensing: present capabilities and future possibilities", *Remote Sensing of Environment*, vol. 35, pp. 243–255, 1991.
2. Estes, J. E., and Mooneyhan, D. W., "Of maps and myths", *Photogrammetric Engineering and Remote Sensing*, vol. 60, pp. 517–524, 1994.
3. Townshend, J. R. G., "Land cover", *International Journal of Remote Sensing*, vol. 13, pp. 1319–1328, 1992.
4. Peddle, D. R., "An empirical comparison of evidential reasoning, linear discriminant analysis and maximum likelihood algorithms for land cover classification", *Canadian Journal of Remote Sensing*, vol. 19, pp. 31–44, 1993.
5. Benediktsson, J. A., Swain, P. H., and Ersoy, O. K., "Neural network approaches versus statistical methods in classification of multisource remote sensing data", *IEEE Transactions on Geoscience and Remote Sensing*, vol. 28, pp. 540–551, 1990.
6. Kanellopoulos, I., Varfis, A., Wilkinson, G. G., and Megier, J., "Land–cover discrimination in SPOT HRV imagery using an artificial neural network - a 20-class experiment", *International Journal of Remote Sensing*, vol. 13, pp. 917–924, 1992.
7. Schalkoff, R. J., *Pattern Recognition: Statistical, Structural and Neural Approaches*, Wiley, New York, 1992.
8. Foody, G. M., Campbell, N. A., Trodd, N. M., and Wood, T. F., "Derivation and applications of probabilistic measures of class membership from the maximum likelihood classification", *Photogrammetric Engineering and Remote Sensing*, vol. 58, pp. 1335–1341, 1992.
9. Maselli, F., Conese, C., and Petkov, L., "Use of probability entropy for the estimation and graphical representation of the accuracy of maximum likelihood classifications", *ISPRS Journal of Photogrammetry and Remote Sensing*, vol. 49, no. 2, pp. 13–20, 1994.
10. Foody, G. M. and Boyd, D. S., (in press), "Observation of temporal variations in environmental gradients and boundaries in tropical West Africa with NOAA AVHRR data", *Proceedings RSS'96: Science and Industry*, Remote Sensing Society, Nottingham.
11. Wang, F., "Improving remote sensing image analysis through fuzzy information representation", *Photogrammetric Engineering and Remote Sensing*, vol. 56, pp. 1163–1169, 1990.
12. Foody, G. M., "Approaches for the production and evaluation of fuzzy land cover classifications from remotely–sensed data", *International Journal of Remote Sensing*, vol. 17, pp. 1317–1340, 1996.
13. Foody, G. M. and Arora, M. K., (in press), "Incorporating mixed pixels in the training, allocation and testing stages of supervised classifications", *Pattern Recognition Letters*.
14. Foody, G. M., Lucas, R. M., Curran, P. J. and Honzak, M., (in press), "Nonlinear mixture modelling without end-members using an artificial neural network", *International Journal of Remote Sensing*.
15. Wang, F. and Hall, G. B., "Fuzzy representation of geographical boundaries in GIS", *International Journal of Geographical Information Systems*, vol. 10, pp. 573–590, 1996.

Geological Mapping Using Multi-Sensor Data: A Comparison of Methods

Paul M. Mather[1], Brandt Tso[1], and Magaly Koch[2]

[1] Department of Geography, The University of Nottingham, Nottingham NG7 2RD, UK
[2] Instituto de Ciencias de la Tierra (Jaume Almera)
Consejo Superior de Investigaciones Cientificas (CSIC)
Lluís Solé i Sabarís s/n, 08028 Barcelona, Spain

Summary. Landsat TM and SIR-C SAR data are used in a comparative test of the performance of a statistical classifier, the maximum likelihood (ML) procedure, and two neural networks, a multi-layer feed-forward network (F-NN) and a Self-Organising Map (SOM). Using spectral features alone, performance (as measured by comparison with unpublished field maps) of all three methods is poor, with overall accuracies of less than 60%. The F-NN performs best. When texture measures are added, overall classification accuracy is improved, with the Grey Level Co-occurrence Matrix (GLCM) approach showing the best result (overall accuracy of almost 70%).

The ML procedure requires less than two minutes' computing time for the 1024^2 test image, whereas the unsupervised learning stage of the SOM required of the order of 70 hours. Design and evaluation of the F-NN was also time-consuming. The cost of using the GCLM procedure to derive texture features is proportional to the number of grey levels in the image (256 in the case of Landsat TM and SIR-C SAR). Reducing the number of grey levels to 64 by means of an equalising algorithm proved to have little effect upon performance.

1. Introduction

Spaceborne Imaging Radar-C (SIR-C) synthetic aperture radar (SAR) data and optical data from the Landsat Thematic Mapper (TM) instrument are used to assess the effectiveness of texture measures in improving the accuracy of lithological classifications, using a test area from the Red Sea Hills, Sudan. Geological details of this area are provided by [1]. Information at optical wavelengths relates to molecular-level interactions that correlate with the mineral composition of the rock and its weathering products. At radar wavelengths, the main contributors to backscatter are surface roughness relative to the radar wavelength, and soil moisture content. Surface roughness, which is related to the composition and weathering characteristics of surface material, is described by the texture of the neighbourhood of a point [2]. The hypothesis to be examined is that textural information may provide additional discriminating capability, as rocks with similar chemical properties may be distinguished on the basis of their surface morphology.

Artificial neural networks (ANN) have been widely used in pattern recognition applications in remote sensing (see [3, 4] for recent reviews). ANN do not require specific assumptions concerning data distributions, in contrast to the more conventional statistical methods. It is generally claimed that the performance of ANN is better than that of conventional statistical classification algorithms. In this work, three classification algorithms, the Gaussian maximum likelihood statistical classifier (ML), a multi-layer feedforward neural network (F-NN), and the Kohonen self-organising feature map (SOM), are used to evaluate their effectiveness in lithological mapping.

Texture can be derived in three main ways:

- modelling, which presumes that the image data possess some property, such as self-similarity [5, 6, 7], or follow a specified frequency distribution, such as log-normal with multiplicative spatial interaction [8]
- joint probabilities of adjacent pixels [9]. From the joint probability information, texture features are extracted in the form of statistical measures. These measures can be computed for different directions and interaction distances.
- analysis of the spatial frequencies in the image. Since different directional structures and scales within the image can be identified in the frequency domain representation, texture information can be extracted through the use of filtering techniques.

2. Study Area and Test Data

The study area is located within the Red Sea Hills of Sudan which is a region of mountainous terrain. The surface lithology is dominated by three main rock types, volcano-sedimentary (VS), granitic batholith (G), and alluvium (A), with a few small patches of intrusive rocks (I). The available remotely-sensed images for the area are SIR-C C- and L-band multi-polarisation SAR data [10] and Landsat Thematic Mapper (TM) imagery. SIR-C C-band HH polarisation and TM optical imagery were chosen because of their better visual quality. The SIR-C image was resampled to a pixel size of 30 m square, and co-registered with the TM image in a UTM projection.

A sub-image of 1024×1024 pixels was selected for this study and 1:100,000 scale unpublished geological maps [11, 12] were used for "ground truth". The classification task is difficult because VS and G rock types cover a large area and are spectrally inhomogeneous, with variations relating to topography and to variations in mineral composition and weathering. The spectra of the intrusive rocks are quite similar to some of the VS and G sequences.

3. Classification Algorithms

The first stage of this work involved the generation of a lithological map of the study area using Landsat TM spectral data alone, using a feed-forward artificial neural network, a self-organising map, and the traditional maximum likelihood statistical classifier. Owing to the considerable overlap of spectral histograms, noted in section 2, the spectral classification was not expected to generate accurate results. Interest lay in determining how well each classifier performed on such difficult data.

3.1 Artificial Neural Networks

Of the range of ANN types and topological structures [13, 14, 15] the neural network architectures implemented in this study are the multi-layer feed-forward neural network (F-NN) and the Self-Organising feature Map (SOM). F-NN are described in many standard texts such as [16]. Several F-NN structures with either one or two hidden layers were investigated in initial experiments, with the learning rate fixed at 0.2. A four-layer network with 36 nodes in hidden layer 1 and 24 nodes in hidden layer 2 was chosen on the basis of its convergence speed and accuracy. Input data were normalised and expressed on the interval 0 – 1. The output data were represented by spread coding.

The SOM ([17, 18, 19]) contains only two layers, the input and output layers. The input layer, or sensory cortex, contains a number of neurons equal to the dimensionality of the data. These features measure the nature of the information which is being used to characterise the data to be classified. The output layer (mapping cortex) is made up of $n \times m$ neurons, lying at the nodes of a regular grid with a grid spacing of unity (measured in terms of Euclidean distance). Values of n and m are generally up to 16. The sensory cortex (input) and the mapping cortex (output) are linked by synaptic weights w_{ij} where the range of i and j corresponds to the number of sensory cortex neurons and mapping cortex neurons, respectively. The weights w_{ij} are continually adjusted during the learning process in order to reflect the sensor cortex properties. The final values of the w_{ij} describe the topological feature space or, in other words, they summarise the patterns presented to the sensory cortex. In summary, each input pixel is presented to the sensory cortex as a set of measurements on a given number of features. The weights map or assign this input pixel to a position in the mapping cortex which it most closely resembles on the basis of its features. The mapping cortex can thus be thought of as a compressed description of the feature space ([4]).

A SOM's learning strategy is based on the competitive (Hebbian) learning procedure and involves the modification of the weights w_{ij} that initially take random values. During the unsupervised learning stage, the learning effect focuses both on a specific neuron and the neighbourhood of that neuron. Thus, the weights w_{ij} associated with neighbouring neurons in the mapping cortex are modified simultaneously. As a result, mapping cortex neurons that are close together will have similar characteristics. At the end of the unsupervised learning stage the weights connecting the sensory cortex to the mapping cortex have been adjusted in order to best describe the nature of the input data. The patterns in the input space are therefore clustered. Schaale and Furrer

([4]) refer to this set of weights as "the codebook". However, if a supervised classification task is to be performed, a second stage of clustering is carried out in order to label the mapping cortex neurons. The supervised labelling concept is based on the concept of the majority vote. The mapping cortex neurons are initially labelled using the training set. Each training pattern is input to the SOM and a corresponding neuron is selected by choosing the minimum Euclidean distance between the given training pattern and weights. If the selected neuron matches the desired output, the corresponding weights are increased. If it does not match, the corresponding weights are decreased.

3.2 Gaussian Maximum Likelihood Classifier

The Gaussian maximum likelihood method (ML) is a well known supervised classification algorithm that is based on the assumption that the population frequency distribution for each class is Gaussian ([20]). Although in practice this assumption is not generally met, the classifier has been widely used, and many examples of its successful application can be found in the literature. Each individual class distribution is characterised by its mean vector and covariance matrix, which are determined from training samples. In order to generate acceptable estimates of these parameters, Swain and Davis ([21]) suggest that the number of training pixels for each class should be at least ten times the input data dimension.

4. Classification Based on Spectral Data

The first classification experiments was based on spectral data using bands 1 to 5 and band 7 of the TM image. For each of the four lithological classes, 800 training samples were selected. One of the problems encountered was the large computing time requirement of the SOM. With a 1024×1024 test image, the unsupervised learning phase of the SOM (6,000 iterations) required 70 hours of processing time on a two-processor workstation. Following the unsupervised training, the SOM was further trained using the LVQ algorithm in order to label each mapping cortex. The number of supervised training iterations was set to 1000. This labelling phase took very little computer time. In comparison, the ML routine took only a few minutes to run on a DEC Alpha workstation.

The best performance, showing an overall accuracy of just over 57% and a kappa coefficient of 0.33, was achieved by the F-NN. This is hardly a high success rate, given the fact that a four-class assignment was being attempted. However, in dealing with such highly confused data, the F-NN does reveal a greater ability to detect class boundaries. The F-NN was therefore chosen as the candidate method to be implemented in the second stage of the experiment, involving the combination of spectral and textural data.

Classification accuracy varied considerably between rock types and between classifiers. User's and producer's accuracy measures show that the ML algorithm identifies only 40.9% of alluvial areas as alluvium; however, all of the areas identified as alluvium are actually in their correct class. The SOM identifies 43.5% of alluvial areas as alluvium, with only 60% user's accuracy, while the F-NN has a producer's accuracy of 77.6% and a user's accuracy of 83.9%. Recognition of alluvial areas is therefore far better using F-NN than either SOM or ML.

The same pattern is apparent in the results for the rock type G, with F-NN achieving higher accuracies than the other two algorithms. However, the producer's accuracy is only 54.6% and there is confusion between granite batholith and other rock types, especially VS. The improvement shown by F-NN over SOM and ML is far less in the case of VS, with F-NN having a producer's accuracy of 57.7% (though the user's accuracy is higher at 81.6%). Confusion between rock types VS and I is apparent, which is not surprising given the overlap of their spectra (section 2).

Rock type I has very low user's accuracies on all three methods, the best being 11.2% in the case of the F-NN. None of the algorithms is capable of resolving the confusion between I, VS and G rock types. In terms of overall performance, therefore, the F-NN is the clear winner, though much of the improvement in classification accuracy is due to the F-NN's ability to recognise the alluvium class which shows an increase in producer's accuracy from 40.9% to 77.6%. For comparison, the same figures for the intrusives class are 46.4% and 50.2%.

5. Classification Using Both Spectral and Textural Information

As in section 4, the spectral data consist of the non-thermal bands of Landsat TM while textural features are derived from C-band SAR data collected by the SIR-C mission. The dimensionality of the TM data was reduced to 3 using the divergence statistic ([20]) as a guide, in order to reduce computational requirements.

Methods for texture characterisation are based on multi-fractals, multiplicative autoregressive random field theory (MAR), second order statistical measures derived from the grey level co-occurrence matrix (GLCM), and frequency domain filtering, respectively. The theoretical background for these four approaches are described in [22]. Since textural features are scale dependent, the choice of window size is critical. Semivariograms were computed for rock types V and G, and a window dimension of 16 selected. Since alluvium is relatively smooth it was thought that texture measures would not have much influence. The area of intrusive rocks is relatively small.

In addition to window size, several other parameters must also be defined for each texture algorithm. For the multi-fractal approach, three weighting factors (qs) were selected as 0, 3, and -4, respectively. For the MAR model, the neighbouring set N was chosen as $\{(1,0),(-1,-1),(-1,0)\}$, and three parameters, σ_u^2, δ_y, and mean entries of θ were used. An inter-pixel distance d of 1.0 and directions of 0°, 45°, 90°, and 135° were used in the computation of the GLCM. Finally, three filters for extracting directional and non-directional texture from the image amplitude spectrum were defined; the directional filters extracted those components of the amplitude spectrum in the range $\pm 22\frac{1}{2}°$ from the horizontal and the vertical respectively, while the non-directional texture consisted of the output from a band-pass filter which extracts frequency components in the range 3 to 7 cycles. This range was selected after some comparative experiments.

The inclusion of texture features shows an improved overall classification accuracy. The features generated by the four texture extraction techniques do not perform equally well, however, nor do they contribute equally to the identification of the individual rock types. The highest classification accuracy is achieved by the combination of spectral and textural features derived by the MAR and GLCM methods, which achieve overall classification accuracies of 68.8 and 69.5, respectively, adding more than 11% to the overall accuracy values in comparison with the use of spectral data alone. The kappa value also rises to 0.47/0.48 respectively. The improvement in classification accuracy resulting from the use of texture features derived from multi-fractal and Fourier methods is less significant, at around 6%.

Rock type A achieves a higher producer's and user's accuracy in comparison with F-NN using spectral features alone for all texture measures except the Fourier-based procedure, which shows a producer's accuracy of 70.9%, which is significantly lower than the value of 77.6 achieved without the textural features. However, user's accuracy for this class exceeds 94% for all texture methods. Producer's accuracy is highest at 81.6% for the multi-fractal approach, with values of 79.9% and 79.7% being reached by MAR and GLCM. Alluvium shows a relatively smooth surface and thus one would not expect any substantial improvements to accrue from the addition of texture features.

For class G, spectral measures achieved a producer's accuracy of 54.6% using the F-NN method; when texture was added this value dropped to 43.3% (multi-fractal), and 42.9% (Fourier), though there was a slight increase to 58% (GLCM) and 61% (MAR). User's accuracy rose from 51.4% to 62.6% and 64.3% in the case of MAR and GLCM respectively, showing that for both of these methods identification of this rock type was improved and confusion lessened, though not by as great an amount as had been expected. However, class VS shows a considerable improvement when texture measures are added. The best result using the six TM bands was 57.7% (producer's) and 81.6% (user's) accuracy. All the texture measures improved on this performance except for the user's accuracy in the case of the multi-fractal and Fourier techniques. The multi-fractal measures improved the producer's accuracy to

75.3%, with a slightly lower user's accuracy of 80.7%, while the other methods attained producer's accuracies of the order of 72 - 74% with user's accuracies in excess of 81.6%.

User's accuracies for class I were all very low (less than 11.2%) using spectral features, and producer's accuracies were between 44 and 50%. The addition of the texture measures generated a marked improvement in the producer's accuracies, especially for GLCM (81%). Although 81% of the intrusive rock pixels were correctly identified, 73.8% of those rocks classified as intrusives were, in fact, some other rock type. While texture features were able to reduce the problem of confusion there is still a substantial difficulty in separating classes I, G and VS.

Classification accuracy in itself may not be the sole determining criterion in the choice of algorithm. Cost is also important. The above results show that the MAR texture algorithm achieves a level of classification accuracy that is almost as high as that derived from the use of GLCM texture features, and at a substantially lower computational cost. Results reported elsewhere ([22]) show that reduction in the number of grey levels from 256 to 64 using an equal probability quantising algorithm ([9]) does not significantly affect the accuracy of the classification.

6. Conclusions

The results of the two experiments demonstrate that the accuracy of automated lithological mapping in semi-arid areas is improved considerably by the use of SAR-based texture measures, though the improvement is not equal in magnitude between the different rock types. Using spectral features alone, classification accuracies between 47% and 57% were obtained, with the multi-layer feed-forward artificial neural network producing the highest accuracy (57.3%). The SOM was computationally very demanding and produced an overall accuracy of 50.4%, or about 3.7% better than the ML algorithm. When texture features were added, classification accuracies rose at best by about 12%, with the MAR and GLCM methods producing the highest accuracies of 68.8% and 69.5% respectively. However, the MAR procedure involves a lower computational overhead. The combination of the F-NN and the MAR texture features is therefore considered to produce an optimum result in terms both of accuracy and computational cost, though if grey-level reduction is employed then the GLCM approach to texture estimation becomes competitive.

Other factors to be considered are:

– SIR-C SAR provides multi-frequency, multi-polarisation data, from which the complete scattering matrix can be derived on a pixel by pixel basis ([10]). The use of multi-frequency polarimetric data can be expected to produce more information about surface roughness than can be obtained from a single waveband ([23]). The performance of texture measures derived from cross-polarised C- and L-band data, which identify surface morphology more clearly ([2]), needs further investigation.

- The rock units selected for this study represent coarse divisions, containing sub-units of differing physical and chemical composition. More refined ground data are needed in order to assess the value of the procedures reported here.

- The effect of differential illumination (for the optical data) and look angle (for radar data) is considerable in the Red Sea Hills area due to the high relative relief. The use of a Digital Elevation Model will provide a means of correcting the data both geometrically and radiometrically, as well as providing the opportunity to introduce geomorphic variables into the analysis.

- The effects of inter-correlation between the spectral and textural features on which the classification is based needs further investigation. Chiuderi ([24]) reports the results of a land cover mapping experiment using a counter-propagation network, showing that, after decorrelating the input channels ([25]), classifier accuracy rose from 72% to 85.6%. The accuracy achieved by a SOM reached almost 98%. The effects of inter-feature correlation appear to be significant in reducing the efficiency of the neural classifiers.

Acknowledgement. The SIR-C radar data was kindly provided by the NASA Jet Propulsion Laboratory, Pasadena California, USA. Mr B. Tso is supported by a scholarship from the Taiwan Government. Computing facilities were provided by the Department of Geography, University of Nottingham.

References

1. Koch, M., *Relationships between Hydrogeological Features and Geomorphic-Tectonic Characteristics of the Red Sea Hills of Sudan based on Space Images.* Unpublished PhD Thesis, Boston University, Boston, Massachusetts, USA, 1993.
2. NASA, *Spaceborne Synthetic Aperture Radar: Current Status and Future Directions* (ed. D. L. Evans). Report to the Committee on Earth Sciences. NASA Technical Memorandum 4679, Washington, D.C. Also available from *http://southport.jpl.nasa.gov/nrc/index.html*, 1995.
3. Paola, J. D. and Schowengerdt, R. A., "A review and analysis of back-propagation neural networks for classification of remotely-sensed multi-spectral images", *International Journal of Remote Sensing*, vol. 16, pp. 3033–3058, 1995.
4. Schaale, M. and Furrer, R., "Land surface classification by neural networks", *International Journal of Remote Sensing*, vol. 16, pp. 3003–3032, 1995.
5. Mandelbrot, B.B., *Fractals: Form, Chance and Dimension*, San Francisco: Freeman, 1977.
6. Mandelbrot, B.B., *The Fractal Geometry of Nature*, San Francisco: Freeman, 1982.

7. Mandelbrot, B.B., "Fractal measures (their infinite moment sequences and dimensions) and multiplicative chaos: Early works and open problems", *Technical Report, Physics Department, IBM Research Center/Mathematics Department, Harvard University*, Cambridge, MA 02138, U.S.A, 1989.

8. Frankot, R. T. and Chellappa, R., "Lognormal random-field models and their applications to radar image synthesis", *IEEE Transactions on Geoscience and Remote Sensing*, vol. GE-25, pp. 195–207, 1987.

9. Haralick, R. M., Shanmugam, K. and Dinstein, I., "Texture feature for image classification", *IEEE Transactions on Systems, Man, and Cybernetics*, vol. SMC-3, pp. 610–621, 1973.

10. NASA, *SIR-C/X-SAR Mission Overview*, JPL Publication 93–29, Jet Propulsion Laboratory, Pasadena, California, 1993.

11. Linnebacher, P. *Geologie und Genese spätproterozoischer Metavulkanite und Plutonite der Red Sea Hills, Sudan - Ein Beitrag zur Rekonstruktion der Geodynamischen Entwicklung des Arabisch-Nubischen Schildes im Späten Präkambrium [Geology and Origin of Late Proterozoic Metavolcanic and Plutonic Assemblages from the Red Sea Hills, Sudan]*. M.Sc. Dissertation, University of Mainz, Germany, 1989.

12. Reischmann, T., *Geologie und Genese spätproterozoischer Vulkanite der Red Sea Hills, Sudan, [Geology and Origin of late Proterozoic Volcanic Rocks in the Red Sea Hills, Sudan]*. Unpublished Ph.D. Thesis, University of Mainz, Germany, 1986.

13. Lippmann, R.P., "An introduction to computing with neural nets", *IEEE ASSP Magazine*, pp. 4–22, 1987.

14. Aleksander, I. and Morton, J., *An Introduction to Neural Computing* (London: Chapman and Hall), 1990.

15. Davalo, E. and Naim, P., *Neural Networks* (Basingstoke: Macmillan), 1991.

16. Beale, R. and Jackson, T., *Neural Computing: An Introduction* (Bristol: Adam Hilger), 1990.

17. Kohonen, T., "Self-organised formation of topologically correct feature maps", *Biological Cybernetics*, vol. 43, pp. 59–69, 1982.

18. Kohonen, T., "An introduction to neural computing", *Neural Networks*, vol. 1, pp. 3–16, 1988.

19. Kohonen, T., "The 'neural' phonetic typewriter", *Computer*, vol. 21, pp. 11–22, 1988.

20. Mather, P.M., *Computer Processing of Remotely-Sensed Images: An Introduction*, Chichester: Wiley, 1987.

21. Swain, P. H. and Davis, S. M., *Remote Sensing: The Quantitative Approach*, New York: McGraw-Hill, 1978.

22. Mather, P. M. and Tso, B., in preparation, Texture extraction: derivations and comparisons.

23. Ulaby, F. T. and Elachi, C., *Radar Polarimetry for Geoscience Applications*, Norwood, MA.: Artech House, Inc, 1990.

24. Chiuderi, A., "Improving the counterpropagation network performances", *Neural Processing Letters*, vol. 2, pp. 27–30, 1995.

25. Gillespie, A. R., Kahle, A. B. and Walker, R. E., "Color enhancement of highly correlated images. I: Decorrelation stretching and HSI contrast stretches", *Remote Sensing of Environment*, vol. 20, pp. 209–235, 1986.

Application of Neural Networks and Order Statistics Filters to Speckle Noise Reduction in Remote Sensing Imaging

E. Kofidis[1], S. Theodoridis[2], C. Kotropoulos[3], and I. Pitas[3]

[1] University of Patras, Department of Computer Engineering, 265 00 Patras, Greece.
[2] University of Athens, Department of Informatics, 157 71 Athens, Greece. e-mail: stheodor@di.uoa.gr
[3] University of Thessaloniki, Department of Informatics, 540 06 Thessaloniki, Greece.

Summary. A novel approach to suppression of speckle noise in remote sensing imaging based on a combination of segmentation and optimum L-filtering is presented. With the aid of a suitable modification of the Learning Vector Quantizer (LVQ) neural network, the image is segmented in regions of (approximately) homogeneous statistics. For each of the regions a minimum mean-squared-error (MMSE) L-filter is designed, by using the histogram of grey levels as an estimate of the parent distribution of the noisy observations and a suitable estimate of the (assumed constant) original signal in the corresponding region. Thus, a bank of L-filters results, with each of them corresponding to and operating on a different image region. Simulation results are presented, which verify the (qualitative and quantitative) superiority of our technique over a number of commonly used speckle filters.

1. Introduction

One of the major problems encountered in Remote Sensing and Ultrasonic Imaging is speckle noise reduction. This type of noise contamination, which is met in all coherent imaging systems, results from the scattering of the transmitted wave from terrain inhomogeneities which are small with respect to the wavelength [4]. A multiplicative model for speckle noise is implied by the fact that the standard deviation is directly proportional to the mean and it has been verified experimentally [9].[1]

The speckle artifact severely degrades the information content of an image and poses difficulties in the image analysis phase. Thus it is desirable to suppress the noise while at the same time retaining the useful information unimpaired. Several algorithms have been proposed aiming at reducing speckle noise in images (e.g., [2, 9, 12, 14, 11]). Since the ultimate goal of any speckle suppression scheme should be the reduction of speckle contrast to enhance the information content of the image, edge and detail preservation are

---

[1] Nevertheless, it must be mentioned that speckle noise is only approximately multiplicative in regions of the object containing fine details that cannot be resolved by the imaging system [18] and the experimental verification in [9] was based only on flat areas of the image. In spite of this, speckle noise is usually modelled as multiplicative in practice.

crucial in a speckle filter along with noise reduction. Thus, spatially-varying filters are required that are also able to deal with the *nonlinear* model governing the degradation process [16].

An important class of adaptive filters is what we call here "segmentation-based filters", that is, filtering processes combining segmentation and (non-adaptive) filters. The underlying idea is that, with the aid of a suitable segmentation algorithm, a statistically non-stationary image can be divided into approximately stationary regions which can, in turn, be processed by filters designed on the basis of the corresponding statistics. Thus, we have a set of filters with each of them corresponding to and operating on a different region of the image, with the various regions being dictated by the segmentation result. In this paper, we report such an approach to speckle suppression employing a modification of the Learning Vector Quantizer neural network at the segmentation stage and non-adaptive minimum mean-squared error (MMSE) L-filters at the filtering stage, designed with the ordering statistical information acquired from the segmentation stage. The proposed filters have been tested on a simulated image containing a bright target in a dark background and the results compare favourably to those produced by a single L-filter designed with the sample statistics of the image considering this as statistically homogeneous. Results of comparison with a number of commonly used speckle filters are also given, which rank our method among the first positions. The noise-smoothing performances of the various filters are compared on the basis of the resulting receiver operating characteristics (ROC's) and an SNR quantity measuring the dispersion of the image pixels in the target and background regions from the corresponding true means. The contrast enhancement effect of the filters is quantitatively assessed through a target contrast measure.

The paper is organised as follows. Our method is presented in detail in section 2. Experimental results are included in section 3, along with a comparison with a number of other well known filtering strategies. Some implementation issues are discussed in section 4, which concludes the paper.

2. Segmentation-Based L-Filtering

In this section we present an adaptive nonlinear approach to speckle suppression in images. The adaptivity of our method comes from the fact that the image is first segmented into regions of different characteristics and each of the resulting regions is processed by a different filter. L-filters are employed to deal with the nonlinear nature of the noise. A number of approaches to the segmentation of speckle images have been reported (e.g., [10]). A recently introduced segmentation technique, that we have adopted in this work, employs a modification of a well known self-organising neural network, the Learning Vector Quantizer (LVQ), based on the L_2 mean which has been shown to be more suitable for speckle images [7].

In the sequel, a brief presentation of LVQ is given followed by the description of its modification, L_2 LVQ, along with a discussion of the need for this modified form. The derivation of the MMSE L-filter for the case of a known constant signal corrupted by noise is included both in the unconstrained and constrained (unbiased) cases. In most cases it is unrealistic to assume that the signal is constant. However, since the filters are matched to specific regions of the image, this simplifying assumption is a good approximation of the reality for practical purposes.

2.1 The Learning Vector Quantizer and its L_2 Mean Based Modification

Learning Vector Quantizer (LVQ) [6] is a self-organising neural network (NN) that belongs to the so-called competitive NN's. It implements a nearest-neighbour classifier using an error correction encoding procedure that could be characterised as a stochastic approximation version of K-means clustering. Let us first present the basic idea. As in the Vector Quantization (VQ) problem, we have a finite set of variable reference vectors (or "code vectors" in the VQ terminology) $\{\mathbf{w}_i(t); \mathbf{w}_i \in \mathcal{R}^N, i = 1, 2, \ldots, p\}$ and a set of training vectors $\mathbf{x}(t) \in \mathcal{R}^N$ where t denotes time and we wish to classify the training vectors into p classes represented by the vectors $\mathbf{w}_i$. These representative vectors are obtained by following an iterative procedure where at each iteration step t the current feature vector $\mathbf{x}(t)$ is compared to all the $\mathbf{w}_i(t)$ and the best-matching $\mathbf{w}_i(t)$ is updated to better comply with $\mathbf{x}(t)$. In this way, in the long run, the different reference vectors tend to become specifically tuned to different domains of the input $\mathbf{x}$. The learning stage of the algorithm is described in the following 4-step procedure:

i. Initialise randomly the reference vectors $\mathbf{w}_i(0)$, $i = 1, 2, \ldots, p$.
ii. At time step t, find the "winner" class c such that:

$$\|\mathbf{x}(t) - \mathbf{w}_c(t)\| = \min_i \{\|\mathbf{x}(t) - \mathbf{w}_i(t)\|\}. \qquad (2.1)$$

iii. Update the winner:

$$\mathbf{w}_c(t + 1) = \mathbf{w}_c(t) + \alpha(t)(\mathbf{x}(t) - \mathbf{w}_c(t)). \qquad (2.2)$$

iv. Repeat steps (ii) and (iii) until convergence.

The gain factor $\alpha(t)$ is a scalar parameter ($0 < \alpha < 1$) which should be a decreasing function of time in order to guarantee the convergence to a unique limit. In the recall procedure, the class with which the input vector $\mathbf{x}(t)$ is most closely associated is determined as in (2.1) where now $\mathbf{w}_i$ is the i-th reference vector after the convergence of the learning procedure.[2]

---

[2] To be precise, we should note that the algorithm described above is the "single-winner " version of LVQ. In its general "multiple-winner" form, step iii above

It is easy to see that eq. (2.2) above is in fact a recursive way of computing the average of the training vectors classified to the class c (this is easily verified by choosing $\alpha(t) = 1/(t+1)$). Thus, after the end of the learning phase, the reference vectors will correspond to the centroids of the associated classes. However, it should be noted that the arithmetic mean approximated by the basic LVQ, described so far, is not the best possible estimator of the mean level in a speckle image. It has been proved [8] that the maximum likelihood estimator of the original noiseless image is the L_2 mean [16] (scaled by $\frac{\sqrt{\pi}}{2}$) of the noisy observations. This result leads us to consider a modification of the standard LVQ algorithm, in which the reference vectors correspond to the L_2 mean instead of the arithmetic mean. The learning and recall parts of the modified algorithm, which we call L_2 LVQ, are exactly analogous to those of the standard LVQ except that the elements of the reference and input vectors are replaced by their squares. This simple modification allows for the computation of the L_2 means providing us at the same time with an algorithm that is proven to be convergent in the mean and in the mean square sense [7].

2.2 MMSE L-Filter Design for a Known Constant Signal Embedded in Noise

The L-filter [1], defined as a linear combination of the input order statistics, has some distinct advantages, making it a right choice for tasks such as the one treated here: it can cope with nonlinear models, it has a relatively simple MMSE design, and furthermore it performs at least as well as, for example, the mean and the median filters, as it includes these filters as special cases [1].

In the sequel, s denotes the constant and known signal, which is corrupted by white[3] noise, independent of s, yielding the noisy observation x. The output of the L-filter of length M is given by:

$$y = \mathbf{a}^T \mathbf{x} \tag{2.3}$$

where $\mathbf{a} = (a_1, a_2, \ldots, a_M)^T$ is the L-filter coefficient vector and $\mathbf{x} = (x_{(1)}, x_{(2)}, \ldots, x_{(M)})^T$ is the vector of the observations arranged in ascending order of magnitude (i.e., order statistics). We will design the optimum in the mean-squared error (MSE) sense L-filter, that is, determine the vector $\mathbf{a}$ minimising $E\{(s-y)^2\}$. By using (2.3) we obtain:

---

involves updating not only the winner vector but its neighbours as well with the neighbourhood defined either in a topological [6] or in a vectorial distance [5] sense.

[3] In fact, speckle noise is locally correlated. Smith et al. [17] argue that, for the observations to be independent, they must belong to different speckle correlation cells. Since our purpose is to apply filters scanning the image in raster fashion, such a recommendation cannot be used directly, thus the whiteness assumption is made to approximate the real situation.

$$E\{(s - y)^2\} = s^2 + \mathbf{a}^T \mathbf{R} \mathbf{a} - 2s\mathbf{a}^T \boldsymbol{\mu} \tag{2.4}$$

where $\mathbf{R} = E\{\mathbf{x}\mathbf{x}^T\}$ is the autocorrelation matrix of the vector of the ordered observations and $\boldsymbol{\mu} = E\{\mathbf{x}\} = (E\{x_{(1)}\}, E\{x_{(2)}\}, \ldots, E\{x_{(M)}\})^T$ is the vector of the expected values of these observations. Setting the derivative of (2.4) with respect to $\mathbf{a}$ equal to zero yields the following expression for the optimum coefficient vector:

$$\mathbf{a} = s\mathbf{R}^{-1}\boldsymbol{\mu} \tag{2.5}$$

It remains to compute the ordering statistics $\boldsymbol{\mu}$ and $\mathbf{R}$. Expressions for the evaluation of these quantities are given in [1] and involve the calculation of the marginal and bivariate probability density functions (pdf's) of the ordered input given its parent distribution:

$$E\{x_{(i)}x_{(j)}\} = \iint xy f_{x_{(i)}x_{(j)}}(x,y)dxdy \quad (i < j) \tag{2.6}$$

$$E\{x_{(i)}\} = \int x f_{x_{(i)}}(x)dx \tag{2.7}$$

where

$$f_{x_{(i)}}(x) = K_i F_x^{i-1}(x)[1 - F_x(x)]^{M-i} f_x(x) \tag{2.8}$$

$$f_{x_{(i)}x_{(j)}}(x,y) = K_{i,j} F_x^{i-1}(x)[F_x(y) - F_x(x)]^{j-i-1}$$
$$\times [1 - F_x(y)]^{M-j} f_x(x) f_x(y) \tag{2.9}$$

and

$$K_i = \frac{M!}{(i-1)!(M-i)!} \tag{2.10}$$

$$K_{i,j} = \frac{M!}{(i-1)!(j-i-1)!(M-j)!} \tag{2.11}$$

Notice that when we are dealing with digital images, the above random variables are of discrete type. Thus, the integrals in eqs. (2.6), (2.7) are in fact discrete sums.

The minimisation of the MSE subject to the constraint that $\mathbf{a}$ provides an unbiased estimate of s, i.e.,

$$s = E\{y\} = \mathbf{a}^T \boldsymbol{\mu}, \tag{2.12}$$

is performed as in the case of additive noise [1] yielding the expression

$$\mathbf{a} = \frac{s\mathbf{R}^{-1}\boldsymbol{\mu}}{\boldsymbol{\mu}^T\mathbf{R}^{-1}\boldsymbol{\mu}} \tag{2.13}$$

for the coefficient vector of the unbiased L-filter.

3. Experimental Results

To test the performance of our method in speckle smoothing and detail preservation, an image consisting of two regions, the target and the background, has been used. For the classification of the image pixels into two groups, we have employed the L_2 LVQ algorithm with parameters $p = 2$ and $N = 49$, trained on a large set of pattern vectors that have been produced by a raster scanning of the image with a 7×7 window. The histograms of the two regions produced by the segmentation have been used as estimates of the parent background and target pdf's, i.e., of the pdf of the random variable x in the background and in the target areas, respectively, for the design of the associated L-filters. Filters of order 3×3 were designed by calculating the ordered statistics from eqs. (2.8)–(2.11) and feeding the results to eqs. (2.6), (2.7) to estimate the quantities $\mathbf{R}$ and μ needed in the computation of the filter coefficients (2.5). The integrals in (2.6) and (2.7) were replaced by sums over the range 0 to 63 since the image's grey levels lie in this interval.

A pair of L-filters have been designed by substituting s in (7) with the L_2 means of the two regions resulted by the segmentation procedure described above.

The arithmetic mean and the median filters have also been used in our comparisons along with a number of well-known speckle filters:

   i. Homomorphic filter [15]
  ii. Frost filter [3, 2]
 iii. Sigma filter [9]
 iv. Variable-length Median filter [12]
  v. Taylor filter [14]

Some detection theoretic performance measures, namely, the probabilities of detection and false alarm, and the receiver operating characteristic have also been used in our comparisons of the filters considered, to allow for numerically comparing their relative performance. The probability of detection corresponding to a threshold chosen so that the probability of false alarm is approximately equal to 10% has been tabulated in table 3.1 for the original image and its processed versions (linear interpolation was used, where necessary, to estimate P_D from its two closest values). The P_D values listed in table 3.1 verify the enhanced detectability obtained by the filters that exploit the segmentation information, compared to their counterparts that are designed with the stationarity assumption. The low probabilities of detection for the median and the mean filters show their inadequacy for this kind of application.

Due to its strong dependency upon the operating point of the detector, the probability of detection P_D, for a fixed probability of false alarm P_F, may be proved an inadequate measure of detection performance. A more reliable figure of merit can be derived by examining the receiver operating characteristic (ROC). A single number that can completely characterise the

Table 3.1. Detection Performance Measures

Method	P_F %	P_D %	Threshold	$\hat{P}_D$ %	Area under ROC
Image	8.198	35.04	22	37.981	0.717116
Thresholding	10.128	38.19	21		
Median	8.737	37.45	21	39.8838	0.743840
	10.60	41.04	20		
Average	9.175	40.907	20	42.592	0.761905
	11.48	45.62	19		
Homomorphic	8.0845	36.91	20	40.88	0.754272
	10.051	40.987	19		
Frost	8.71	43.73	17	46.021	0.77281
	12.3	50.11	16		
Sigma	9.17	40.845	20	42.56	0.761576
	11.476	45.61	19		
V. L. Median	8.78	37.44	21	39.6336	0.731703
	10.68	40.854	20		
Taylor	9.175	40.9	20	42.592	0.761882
	11.48	45.61	19		
L-filter	8.674	38.8	21	41.77	0.758334
	10.7	43.343	20		
L-filter pair	7.838	40.3667	18	44.6027	0.764672
	10.406	45.3982	17		

whole ROC is the area under this curve and is included in table 3.1. The comparison with respect to this figure of merit is again seen to be favourable for our method.

We have also compared the various filtering strategies from the viewpoint of the dispersion of the background and target pixels from the corresponding true sample means relatively to the dispersion in the original image. A measure of this relative dispersion, that could be called a signal-to-noise ratio, is defined for the target area as

$$\text{SNR}_T = \frac{\sum_{\text{target}} (x_i - m_T)^2}{\sum_{\text{target}} (\hat{x}_i - m_T)^2} \tag{3.1}$$

where x_i and $\hat{x}_i$ denote the values of the original and the filtered image, respectively, and m_T corresponds to the average level in the target that is estimated from the original image on the basis of our a-priori knowledge of the target position, shape, and dimensions. The background SNR, SNR_B, is similarly defined. Table 3.2 summarises the SNR values (in decibels) for our set of processed images.

The results presented thus far, demonstrate that our method outperforms all of the filters considered except for the Frost filter, which attains significantly higher values for the ROC area and the SNR compared to the segmentation-based approach. Nevertheless, these higher figures of merit for the Frost filter are at the cost of lower target contrast and an amount of

Table 3.2. Relative dispersion in the target and background areas

Method	SNR$_B$ (dB)	SNR$_T$ (dB)
Median	0.914986	0.885867
Arithmetic Mean	1.55226	1.5066
Homomorphic	1.56488	1.50149
Frost	4.59596	4.02855
Sigma	1.54537	1.49875
V. L. Median	0.812148	0.613674
Taylor	1.55216	1.5066
L-filter	1.09915	1.04087
L-filter pair	2.28008	2.1488

blurring. This could be expected since, as noted in [11], Frost's filter cannot adequately smooth homogeneous areas and preserve heterogeneous areas at the same time. A quantitative verification of this point is provided by the following measure of target contrast

$$C = \frac{m_T - m_B}{m_T + m_B} \qquad (3.2)$$

where, as before, m_T and m_B denote the average levels in the target and the background, respectively. The contrast values for the original as well as the filtered images are tabulated in table 3.3. Note that the Frost filter yields the lowest target contrast among all the filters studied here, with the highest value obtained through our method.

Table 3.3. Target contrast

Method/Image	Contrast
Original	0.230441
Median	0.230212
Average	0.230541
Homomorphic	0.231008
Frost	0.187298
Sigma	0.230571
V. L. Median	0.228513
Taylor	0.230541
L-filter	0.238434
L-filter pair	0.238952

4. Conclusions

We have presented a method for the suppression of speckle noise in remote-sensing imagery based on the idea of segmenting an image into stationary sub-images prior to processing each of them with a filter that is designed to be optimal for each particular sub-image on the basis of the (statistical) information provided by the segmentation. Our method employs a modification of the LVQ algorithm based on the L_2 mean as a means of segmentation, while its filtering stage uses L-filters that are optimal in the MSE sense. The simulation results verified the superiority of our approach to the single L-filter as well as to a number of commonly used speckle filters.

The L_2 LVQ training and the subsequent computation of more than one filters increase the computational complexity of our method to more than twice the computation needed in a conventional approach. However, this need not to be a problem if the processing can be done off-line. Moreover, the generalisation capability of the LVQ NN, which has already been verified by simulations on SAR images [13], could be exploited to considerably reduce the computational load by using a single fixed network for segmentation, trained on a sufficiently large and representative sample of images.

References

1. A. C. Bovik, T. S. Huang, and D. C. Munson, "A Generalisation of Median Filtering Using Linear Combinations of Order Statistics", *IEEE Transactions Acoustics, Speech, and Signal Processing*, vol. 31, no. 6, December 1983, pp. 1342–1350.
2. V. S. Frost, J. A. Stiles, K. S. Shanmugam, and J. C. Holtzman, "A Model For Radar Images and Its Application to Adaptive Digital Filtering of Multiplicative Noise", *IEEE Transactions Pattern Analysis and Machine Intelligence*, vol. 4, no. 2, March 1982, pp. 157–165.
3. V. S. Frost, J. A. Stiles, K. S. Shanmugam, J. C. Holtzman, and S. A. Smith, "An Adaptive Filter for Smoothing Noisy Radar Images", *Proceedings IEEE*, vol. 69, no. 1, Jan. 1981, pp. 133–135.
4. J. W. Goodman, "Some Fundamental Properties of Speckle", *J. Optical Society of America*, vol. 66, no. 11, Nov. 1976, pp. 1145–1150.
5. J. A. Kangas, T. K. Kohonen, and J. T. Laaksonen, "Variants of Self-Organizing Maps", *IEEE Transactions Neural Networks*, vol. 1, no. 1, March 1990, pp. 93–99.
6. T. Kohonen, "The Self-Organizing Map", *Proceedings IEEE*, vol. 78, no. 9, September 1990, pp. 1464–1480.
7. C. Kotropoulos, X. Magnisalis, I. Pitas, and M. G. Strintzis, "Nonlinear Ultrasonic Image Processing based on Signal-Adaptive Filters and Self-Organizing Neural Networks", *IEEE Transactions Image Processing*, vol. 3, no. 1, Jan. 1994, pp. 65–77.
8. C. Kotropoulos and I. Pitas, "Optimum Nonlinear Signal Detection and Estimation in the Presence of Ultrasonic Speckle", *Ultrasonic Imaging*, vol. 14, 1992, pp. 249–275.
9. J.-S. Lee, "A Simple Speckle Smoothing Algorithm for Synthetic Aperture Radar Images", *IEEE Transactions Systems, Man, and Cybernetics*, vol. 13, no. 1, Jan/Feb. 1983, pp. 85–89.
10. J.-S. Lee and I. Jurkevich, "Segmentation of SAR Images", *IEEE Transactions Geoscience and Remote Sensing*, vol. 27, no. 6, November 1989, pp. 674–680.

11. A. Lopes, R. Touzi, and E. Nezry, "Adaptive Speckle Filters and Scene Heterogeneity", *IEEE Transactions Geoscience and Remote Sensing,* vol. 28, no. 6, November 1990, pp. 992–1000.
12. T. Loupas, W. N. McDicken, and P. L. Allan, "An Adaptive Weighted Median Filter for Speckle Suppression in Medical Ultrasonic Images", *IEEE Transactions Circuits and Systems,* vol. 36, no. 1, January 1989, pp. 129–135.
13. S. P. Luttrel, "Image Compression Using a Multilayer Neural Network", *Pattern Recognition Letters,* vol. 10, July 1989, pp. 1–7.
14. C. R. Moloney and M. E. Jernigan, "Nonlinear Adaptive Restoration of Images with Multiplicative Noise", *Proceedings ICASSP '89,* pp. 1433–1436.
15. A. V. Oppenheim, R. W. Schafer, and T. G. Stockham, Jr., "Nonlinear Filtering of Multiplied and Convolved Signals", *Proceedings IEEE,* vol. 56, August 1968, pp. 1264–1291.
16. I. Pitas and A. N. Venetsanopoulos, *Nonlinear Digital Filters: Principles and Applications,* Kluwer Academic Publishers, Hingham MA, 1990.
17. S. W. Smith, R. F. Wagner, J. M. Sandrik, and H. Lopez, "Low Contrast Detectability and Contrast/Detail Analysis in Medical Ultrasound", *IEEE Transactions Sonics and Ultrasonics,* vol. 30, no. 3, May 1983, pp. 164–173.
18. M. Tur, K. C. Chin, and J. W. Goodman, "When is Speckle Noise Multiplicative?", *Applied Optics,* vol. 21, no. 7, April 1982, pp. 1157–1159.

Neural Nets and Multichannel Image Processing Applications

Vassilis Gaganis, Michael Zervakis, and Manolis Christodoulou

Department of Electrical & Computer Engineering.
Technical University of Crete, 73100 Chania, Crete, GREECE.
e-mail: {gaganis,michalis}@ced2.tuc.gr

Summary. The application of neural network technology to multichannel image processing is presented in this paper. Topics such as image restoration, segmentation, transformation and compression are discussed, covering a wide range of image processing and analysis areas. The problems are converted to optimisation or interpolation problems through the appropriate mathematical interface. This alternative interpretation enables the application of well-known neural network structures, perfectly suited for such purposes due to the accuracy and high speed of computation. The proposed framework handles multichannel data in a compact form and, thus, it is directly applicable to remote sensing.

1. Introduction

In this paper, we present state-of-the-art applications of neural networks (NN) to multichannel image processing and provide the framework for similar potential applications to remote sensing. We focus on the area of image restoration and discuss emerging developments in segmentation, compression and transformations. Thus, NN cover direct processing and analysis techniques, but also provide a means of data transformation and processing in different domains.

1.1 Static Neural Networks

A Back-Propagation NN (figure 1.1 consists of one input layer, one output layer and a number of hidden layers of neurons [1, 2]. The number of neurons of the input and output layers is equal to the dimensionality of the input and output vectors of the unknown function to be learned respectively. All outputs of a single layer are multiplied by the corresponding interconnection strength (weight) and summed up in the next layer, where they also pass through a non-linear activation function. Most often the activation is implemented using the sigmoid function :

$$s(x) = \frac{k}{1 + e^{-lx}} + \lambda \tag{1.1}$$

where k and λ determine the sigmoid range and offset, while $l > 0$ determines its slope.

The same rule is used to propagate the signal towards the output layer. Denoting by $\mathbf{W}_i$ the weight matrix following the i layer, the network behaviour may be described by

$$\mathbf{y} = \mathbf{S}(\mathbf{W}_n \mathbf{S}(\mathbf{W}_{n-1} \dots \mathbf{W}_1 \mathbf{x}) \dots) \qquad (1.2)$$

where $\mathbf{S}(x) = [s(x_1), s(x_2), \dots, s(x_n)]^t$.

The "training" procedure aims at adjusting the weight values in such a way that the error between the NN output and the desirable output for a given pair of input-output data becomes as small as possible. Hence it requires the minimisation of

$$E = \left\{ \sum_{i=1}^{p} \{ \mathbf{S}(\mathbf{W}_n \mathbf{S}(\mathbf{W}_{n-1} \dots \mathbf{W}_1 \mathbf{x}_i) \dots) - \mathbf{y}_i \}^2 \right\} \qquad (1.3)$$

For this purpose, simple optimisation techniques, like the gradient descent method are employed, providing weight update rules that force the weights towards a direction opposite to the error function gradient. The update rule for each weight matrix is given by

$$\frac{dw_{ij}}{dt} = -\gamma \frac{\partial E}{\partial w_{ij}} \qquad (1.4)$$

where γ is a positive gain factor.

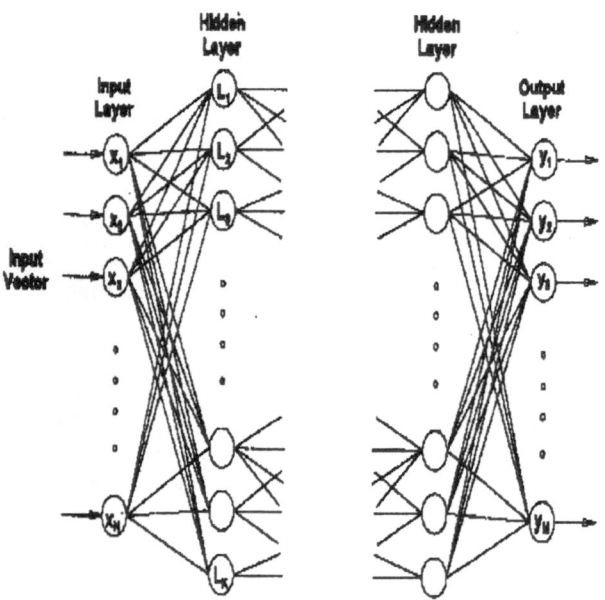

Fig. 1.1. Back-Propagation neural network configuration

1.2 Dynamic (Hopfield) Neural Networks

Another family of NN structure is derived by allowing feedback connections, leading to the notion of dynamic networks [1]. Once an input signal is constantly applied, neuron states evolve in time until equilibrium is obtained.

The dynamic equation governing the system's behaviour can be written in the following form

$$v_i(k+1) = s\left(\sum_{j=1}^{n} w_{ij} u_j(k) + I_i\right) \qquad (1.5)$$

where w_{ij} stands for the interconnection strength between neurons i and j, while I_i denotes the constant input applied.

The main result, due to Hopfield [3], concerns system stability and energy reduction (figure 1.2). If $\mathbf{W}$ is a symmetric matrix with $w_{ii} > 0$, the network is guaranteed to converge to some equilibrium point, while the energy function

$$E = -\frac{1}{2}\mathbf{v}^t\mathbf{W}\mathbf{v} - \mathbf{I}^t\mathbf{v} \qquad (1.6)$$

is non-increasing throughout the NN transient response. Obviously, by selecting appropriate values for both weights and inputs, the Hopfield network can be used for solving optimisation problems by mapping the objective function to be minimised onto NN internal energy. The main advantage of such an approach is the solution speed which depends exclusively on the time constants of the electrical elements used for the hardware implementation of this structure.

2. Multichannel Image Restoration

2.1 Problem Definition

Consider the formation of a colour image, consisting of K channels (RGB, HSI, etc). Let $\mathbf{f}^k$ and $\mathbf{g}^k$ denote, respectively, for the kth channel the original and degraded image vectors, $\mathbf{n}^k$ denote the noise vector, and the matrix $\mathbf{H}_{kk}$ represent the degradation matrix, consisting of point spread functions (PSF). For an $N \times N$ image, the image and noise vectors dimensionality in each channel is equal to N^2, whereas each PSF operator is a $N^2 \times N^2$ matrix. Writing the original image vector in the form $\mathbf{f} = [\mathbf{f}^{1t}\mathbf{f}^{2t}\ldots\mathbf{f}^{Kt}]$ and assuming similar forms for the degraded image and noise vectors, the overall degradation matrix is written in block form as

$$\mathbf{H} = \begin{bmatrix} \mathbf{H}_{11} & \mathbf{H}_{12} & \cdots & \mathbf{H}_{1K} \\ \mathbf{H}_{21} & \mathbf{H}_{22} & \cdots & \mathbf{H}_{2K} \\ \vdots & \vdots & \ddots & \vdots \\ \mathbf{H}_{K1} & \mathbf{H}_{K2} & \cdots & \mathbf{H}_{KK} \end{bmatrix} \qquad (2.1)$$

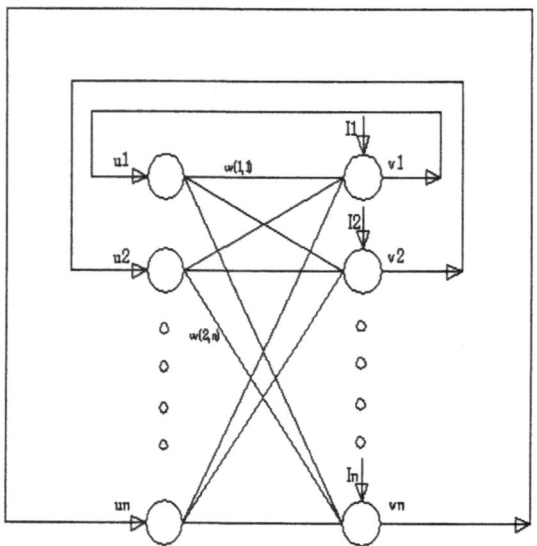

Fig. 1.2. Hopfield dynamic neural network configuration

This notation leads to the overall image formation model [4, 5, 6]

$$\mathbf{g} = \mathbf{Hf} + \mathbf{n} \qquad (2.2)$$

This formulation is general enough to cover other applications, including remote sensing.

The multichannel image restoration problem can be interpreted as a generalised MAP approach, as shown in [4], leading to an estimate given by

$$\hat{\mathbf{f}}(\alpha) = \arg\{\min_f Q(\alpha, \mathbf{f})\} = \arg\{\min_f \{R_n(\mathbf{g} - \mathbf{Hf}) + \alpha R_f(\mathbf{Cf})\}\} \qquad (2.3)$$

where α is referred to as the *regularisation parameter* and $\mathbf{C}$ is the *regularising operator*. This operator can take the form of the 3-D Laplacian filter, or the 2-D Laplacian filter applied independently on the different channels of the image. In addition, a channel adaptive form $\mathbf{C}$ is proposed in [6], where one colour channel affects another channel according to the similarity of the overall brightness in these channels. The $R_f(.)$ and $R_n(.)$ functionals, which are referred to as the "noise" and "signal" objective functionals are expressed in terms of the kernel functions $r_f(.)$ and $r_n(.)$ respectively

$$R_n(\mathbf{g} - \mathbf{Hf}) = \sum_{m=1}^{KN^2} r_n \left(g_m - \sum_{j=1}^{KN^2} H_{mj} f_j \right) \qquad (2.4)$$

and

$$R_f(\mathbf{Cf}) = \sum_{m=1}^{KN^2} r_f \left(\sum_{j=1}^{KN^2} C_{mj} f_j \right) \qquad (2.5)$$

The kernel functions are usually defined by their derivatives $\phi_n(.)$ and $\phi_f(.)$, respectively, referred to as the *influence functions*. The influence functions in many applications assume the form

$$\phi(x) = \tau \tanh(\gamma x) \tag{2.6}$$

Note that the general form of the optimisation problem (2.3) can also be obtained from the Tichonov-Miller approach [5, 7].

Non-quadratic kernel functions $r_n(.)$ and $r_f(.)$ lead to robust metrics, where large deviations from zero are penalised less than by a quadratic metric. In this case robust estimates can be obtained using iterative schemes, such as the gradient descent and the Gauss-Newton algorithms.

Taking the special case of quadratic functionals

$$R_n(\mathbf{x}) = \mathbf{x}^t \mathbf{L_n} \mathbf{x} \qquad R_f(\mathbf{x}) = \mathbf{x}^t \mathbf{L_f} \mathbf{x} \tag{2.7}$$

where $\mathbf{L_n}$ and $\mathbf{L_f}$ are diagonal weight matrices, the solution of the MAP criterion can be obtained analytically as the linear estimator

$$\hat{\mathbf{f}}(\alpha) = \left[\mathbf{H^t L_n H} + \alpha \mathbf{C^t L_f C} \right]^{-1} \mathbf{H^t L_n g} \tag{2.8}$$

2.2 Neural Networks in Multichannel Image Restoration

Consider the case of quadratic norms in (2.3), which is equivalent to the optimisation problem

$$\hat{\mathbf{f}}(\alpha) = \arg \left\{ \min_f \left\{ \mathbf{x}^t \mathbf{A} \mathbf{x} - \mathbf{b}^t \mathbf{x} + c \right\} \right\} \tag{2.9}$$

where $\mathbf{A} = \mathbf{H^t L_n H} + \alpha \mathbf{C^t L_f C}$, $\mathbf{b} = 2\mathbf{g^t L_n H}$ and $c = \mathbf{g^t L_n g}$. The constant term c can be omitted since it does not affect the optimisation result.

Consider a Hopfield NN consisting of KN^2 neurons, each representing the value of one channel of one pixel. Setting the neuron inputs equal to $I_i = b_i/A_{ii}$, the weights equal to $W_{ij} = -A_{ij}/A_{ii}$ and applying the update rule

$$x_i(k+1) = x_i(k) + \sum_{i=1}^{KN^2} w_{ij} x_j(k) + I_i \tag{2.10}$$

the NN implements the Gauss-Seidel algorithm for solving the linear system of restoration equations

$$\mathbf{A}\mathbf{x} = \mathbf{b} \tag{2.11}$$

which also implements the Gauss-Seidel iterative algorithm for the solution of the optimisation problem (2.3) [8]. The Gauss-Seidel iterative algorithm converges if $\mathbf{A} > 0$ and $A_{ii} > 0$ which can be easily proved to be the case in the formulation of (2.9). The presented scheme is an asynchronous one [2], that is each neuron is updated sequentially with a random order. Obviously such an algorithm does not fully exploit the NN capabilities, and thus a synchronous update scheme is required.

Selecting an arbitrary matrix $\mathbf{G} = \text{diag}\{\epsilon_1, \epsilon_2, \ldots, \epsilon_{KN^2}\}$ such that $\epsilon_i > 1/2 \sum_{j=1}^{j=KN^2} \|A_{ij}\|$, setting the weights matrix and the network inputs respectively equal to

$$\mathbf{W} = -\mathbf{G}^{-1}\mathbf{A} \quad \mathbf{I} = \mathbf{G}^{-1}\mathbf{b} \tag{2.12}$$

and using update rule (2.10) in a synchronous mode, that is

$$\mathbf{x}(k+1) = \mathbf{x}(k) + \mathbf{W}\mathbf{x} + \mathbf{I} \tag{2.13}$$

the Jacobi algorithm is implemented by means of dynamic NNs. Defining $\mathbf{H} = \mathbf{G} - \mathbf{A}$ we can easily prove that $\|\mathbf{G}^{-1}\mathbf{H}\| < 1$ and thus the error equation

$$e(k) = \mathbf{x}(k) - \mathbf{x}^\star = (\mathbf{G}^{-1}\mathbf{H})^k e(0) \tag{2.14}$$

where $\mathbf{x}^\star$ stands for the exact solution, converges to zero exponentially fast.

Considering the more general robust case in (2.3), the energy function gradient is given by

$$\nabla_f Q = \mathbf{H}^t \phi_n(\mathbf{H}f - \mathbf{g}) + \alpha \mathbf{C}^t \phi_f(\mathbf{C}f) \tag{2.15}$$

A gradient descent algorithm leads to the following iterative scheme

$$\frac{d\mathbf{f}}{dt} = -\Lambda \nabla_f Q = -\Lambda \mathbf{H}^t \phi_n(\mathbf{H}f - \mathbf{g}) - \alpha \Lambda \mathbf{C}^t \phi_f(\mathbf{C}f) \tag{2.16}$$

where $\Lambda = \{d_1, d_2, \ldots, d_{3N^2}\}$ is a positive definite diagonal gain matrix. The algorithm above can be written element-wise in the simpler form

$$\frac{df_i}{dt} = -d_i \sum_k h_{ki}\phi_n(e_k) - \alpha d_i \sum_k c_{ki}\phi_f(s_k) \tag{2.17}$$

where $e_k = \sum_j (h_{kj}f_j - g_k)$ and $s_k = \sum_j c_{kj}f_j$. Simplifying further we obtain

$$\frac{df_i}{dt} = -d_i \sum_k h_{ki}\psi_n(\mathbf{f}) - \alpha d_i \sum_k c_{ki}\psi_f(\mathbf{f}) \tag{2.18}$$

Note that (2.18) is a weighted linear sum of $\psi_n(\mathbf{f})$ and $\psi_f(\mathbf{f})$ and thus can be mapped onto a Hopfield-like NN, with a single neuron state but two activation functions per neuron performing the calculation of $\psi_n(\mathbf{f})$, $\psi_f(\mathbf{f})$ and transmitting both outputs to other neurons k through the weights $-d_i h_{ki}$ and $-\alpha d_i c_{ki}$ respectively [9]. The important result is that the scheme described is a synchronous one which evolves and stabilises rapidly. Other image restoration techniques concerning modified neural networks which do not satisfy Hopfield convergence criteria but also guarantee energy reduction can be found in [10] and [11].

In case matrix $\mathbf{H}$ cannot be estimated accurately, the above techniques cannot be applied, since the degradation model is assumed to be completely unknown. However, given a large number of original and degraded image pairs, we can employ a static NN to interpolate the inverse effect of the blurring function and the additive noise [12]. Actually, since the support of the blurring function is quite small compared to the image dimensions, we may train a static neural network in a way that given a degraded image window, the network outputs the original colour of the centred pixel. For a K-channel image and a $M \times M$ window, the Back-Propagation network has a $K \times M^2$ dimensional input vector, and a K-dimensional output vector. In both the input and output vectors one neuron is used per pixel and per channel.

Given a large number of training pairs, the NN is trained using standard gradient descent. Obviously, all training pairs should be obtained from images taken by the same source, so as to maintain the same blurring function and additive noise properties. The training procedure results in a NN which can produce the original colour of a pixel given a surrounding window of the degraded image. In this form, the NN essentially learns the inverse of the blurring matrix $\mathbf{H}$.

3. Multichannel Image Segmentation

Our objective is to classify N pixels of P features (levels) per channel, among M clusters such that the assignment of the pixels minimises a criterion function measuring the dissimilarity of all pixels within the same cluster [13]. The colour of a pixel is described by K P-dimensional feature vectors $\mathbf{X}^{(1)}$, $\mathbf{X}^{(2)}$, ..., $\mathbf{X}^{(K)}$, whose entries are all equal to zero, but one which is equal to the unity. A possible distribution can be fully described by a set of variables V_{kl} with $1 \leq k \leq N$ and $1 \leq l \leq M$, where $V_{kl} = 1$ denotes that pixel k belongs to cluster l. For a distribution to be valid we require that

$$\forall k \quad \sum_{i=1}^{M} V_{ki} = 1 \ , \qquad \forall i,j \leq N, \ i \neq j \ \ V_{ki}V_{kj} = 0 \tag{3.1}$$

Let us define the generalised distance measure

$$R_{kl}^{(i)} = \|\mathbf{X}_k^{(i)} - \bar{\mathbf{X}}_l^{(i)}\|^2_{\mathbf{A}_l^{(i)}} = \left(\mathbf{X}_k^{(i)} - \bar{\mathbf{X}}_l^{(i)}\right)^t \mathbf{A}_l^{(i)} \left(\mathbf{X}_k^{(i)} - \bar{\mathbf{X}}_l^{(i)}\right) \tag{3.2}$$

where $\bar{\mathbf{X}}_l^{(i)}$ stands for the P-dimensional centroid of channel i of class l, that is

$$\bar{\mathbf{X}}_l^{(i)} = \frac{\sum_{k=1}^{N} \mathbf{X}_k^{(i)} V_{kl}}{n_l} \tag{3.3}$$

and n_l is the number of pixels in class l. The quantity $R_{kl}^{(i)}$ in (3.2) measures the dissimilarity of each pixel with respect to the rest pixels of the same channel within the same cluster. The matrix $\mathbf{A}_l^{(i)}$ can be a unitary matrix, leading to a Euclidean distance measure, or it can be set to the inverse cluster covariance matrix. In the latter case, the smaller the correlation between the candidate pixel and the centroid, the larger the contribution to the total dissimilarity measure.

$R_{kl}^{(i)}$ should be included in the criterion only if pixel k belongs to cluster l, which can be expressed by the value of the quantity $R_{kl}^{(i)} V_{kl}^2$. Thus, summing together all distortions, we construct the criterion to be minimised

$$E = \frac{1}{2} \sum_{k=1}^{N} \sum_{l=1}^{M} \left(\sum_{i=1}^{K} R_{kl}^{(i)} V_{kl}^2 \right) \tag{3.4}$$

To force the NN evolve towards the minimisation of the criterion E, we require that it is governed by the motion equations

$$\frac{\partial U_{kl}}{\partial t} = -\frac{\partial E}{\partial V_{kl}} \tag{3.5}$$

where U_{kl} is the state of the ith neuron. The motion equations above lead to the gradient descent algorithm. For the case of the criterion of (3.4), we get a set of equations for the neuron dynamics given by

$$\frac{dU_{kl}}{dt} = -R_{kl} V_{kl} \tag{3.6}$$

where the distance measure R_{kl} is an implicit function of the state V_{kl} and is changing at each iteration with the new estimation of the weighting matrix $\mathbf{A}_l$.

To ensure that a valid distribution is obtained we apply the the following neuron activation function

$$\begin{aligned} V_{kn}(k+1) &= 1 \quad \text{if } U_{kn} = \max\{U_{kl}(t),\ \forall l\} \\ V_{kl}(t+1) &= 0 \quad \text{otherwise} \end{aligned} \tag{3.7}$$

which is the Winner Take All model, as it is well-known in NN technology [17].

4. Multichannel Image Compression

Towards the compression of gray-level images, we can split the image into small (usually 8×8) patches in order to take advantage of the benefits of vector quantisation. Each patch is organised into a M-dimensional vector.

Consider a set of n M-dimensional vectors $\mathbf{x}_i$ and an orthonormal $P \times M$ matrix $\mathbf{C}$ where $P << M$. If $\mathbf{C}$ is such that $\mathbf{C}^t \mathbf{C} \mathbf{x} \approx \mathbf{x}$, $\forall \mathbf{x}$ we can store the set $\mathbf{y} = \mathbf{C}\mathbf{x}$ together with matrix $\mathbf{C}$, and use the transpose matrix to obtain the original vectors $\mathbf{x} \approx \mathbf{C}^t \mathbf{y}$, thus achieving significant compression of the original set. In fact the original N-dimensional vector space is projected to a P-dimensional subspace whose basis consists of the principal components of the original space [14, 15], as it can be seen in figure 4.1. The compressed data set requires storage space equal to $n \times P$ for the compressed vector versions and $M \times P$ for the compression matrix $\mathbf{C}$, compared to $n \times M$ for the original set. The larger the ratio M/P, the larger the compression ratio

$$\frac{n \times M - (n \times P + M \times P)}{n \times M}$$

For the case of multichannel image compression, we consider K channels (frames) split into small patches (usually 8×8 pixels) and collect all patch pixels in MK-dimensional channel vectors, organised in K-dimensional entries. In this case the $MK \times MK$ matrix $\mathbf{C}$ is partitioned in $K \times K$-dimensional sub-matrices. At this point, we employ NN techniques to calculate the projection matrix $\mathbf{C}$.

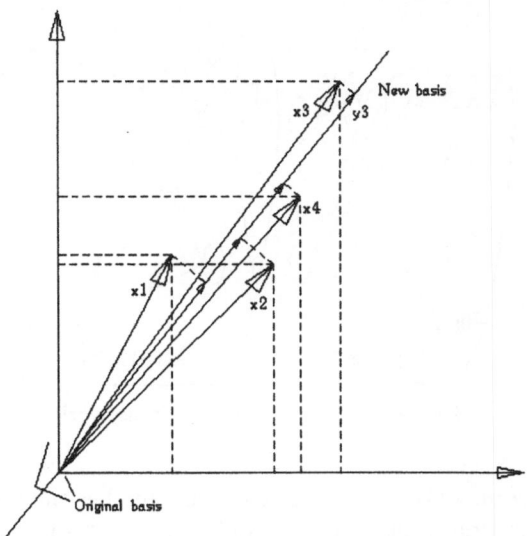

Fig. 4.1. High-dimensional space projection

Consider the NN structure shown in figure 4.2 [16]. The first weight-layer corresponds to the transmitter part, while the second weight-layer corresponds to the receiver part, where the compressed data are transformed back to the original space. In order to set up the weights, or the elements of matrix $\mathbf{C}$, the original channel vectors are applied one after another to the input NN, and at each iteration the matrix $\mathbf{C}$ elements (weights) are adjusted in a way that the error function

$$\sum_{i=1}^{N} \|\mathbf{C}^t \mathbf{C} \mathbf{x}_i - \mathbf{x}_i\|^2 \tag{4.1}$$

is minimised, under the constraint that matrix $\mathbf{C}$ remains orthonormal. This goal can be achieved through the use of a modified steepest descent algorithm, where the update vector is given by [16]

$$\Delta\mathbf{C} = \gamma \left(\mathbf{x}\mathbf{r}^t + \mathbf{r}\mathbf{x}^t\right) \mathbf{C} \tag{4.2}$$

where γ is the learning rate, and $\mathbf{r}$ is the projection residual defined by $\mathbf{C}^t\mathbf{y} - \mathbf{x} = (\mathbf{C}^t\mathbf{C} - \mathbf{I})\,\mathbf{x}$. The modified algorithm actually rotates matrix $\mathbf{C}$ columns so that the orthonormality condition remains unattached.

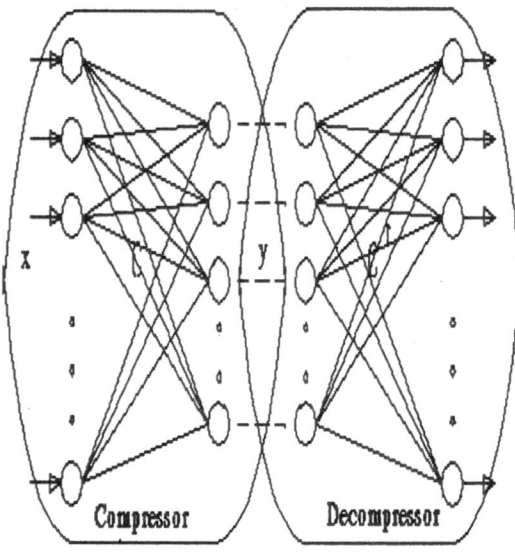

Fig. 4.2. Image compression network

A second NN may also be used to compress further the residuals obtained by the first projection, which are expected to form a "noise subspace". In this case, the estimate of the original image is obtained by summing the outputs of both output NNs.

5. Multichannel Image Transformations

The discrete Fourier transform (DFT) is encountered frequently in image processing. If the DFT calculation can be mapped onto an optimisation problem, NNs can be employed leading to a solution method that operates on entire vectors simultaneously and reaches a solution in a time determined by RC time constants, rather than by algorithmic time complexity. Other transforms can be also computed in a similar scenario. We further discuss the discrete Hartley transform (DHT) due to its simplicity and its strong relation to DFT.

The DHT of a set of sampled data and its inverse are defined as follows [18]:

$$u(v) = \frac{1}{N}\sum_{\tau=0}^{N-1} b(\tau)\mathrm{cas}\left(\frac{2\pi v\tau}{N}\right) \qquad b(\tau) = \sum_{v=0}^{N-1} u(v)\mathrm{cas}\left(\frac{2\pi v\tau}{N}\right) \qquad (5.1)$$

where N is the number of samples (pixels), $\mathbf{b}$ is the sampled data, $\mathbf{u}$ is the DHT of $\mathbf{b}$, and $cas(x) = cos(x) + sin(x)$. In matrix format, the inverse DHT can be written in the form

$$\mathbf{b} = \mathbf{Du} \qquad (5.2)$$

The DFT can be obtained by means of the DHT using

$$f(v) = E(v) + jO(v) = \frac{u(v) + u(N-v)}{2} + j\frac{-u(v) + u(N-v)}{2} \qquad (5.3)$$

Consider the linear programming neural network of figure 5.1, which is proven to minimise the energy function [19]

$$E = \sum_i a_i u_i + \sum_j F\left(\sum_i D_{ji}u_i - b_j\right) + \sum_i \frac{1}{R_i}\int_0^{u_i} g^{-1}(u)du \qquad (5.4)$$

where f is the constraint amplifier function, F is the indefinite integral of f, g is the signal amplifier function, u_i is the output voltage of signal amplifier i, $\mathbf{a},\mathbf{b}$ are the input currents to signal and constraint planes and R_i is the resistance at input of signal amplifier i [19, 20].

Setting $\mathbf{a}=0$, $\mathbf{b}$ equal to the sampled data series, $g(u) = \beta u$ and $f(z) = \alpha z$ giving $F(z) = (\alpha/2)z^2$, the energy function (5.4) becomes equal to

$$E = \frac{\alpha}{2}\|\mathbf{Du} - \mathbf{b}\|^2 + \frac{1}{\beta R}\|\mathbf{u}\|^2 \qquad (5.5)$$

where all R_i's are set to the same value R [21]. The first term reaches its minimum value as soon as $\mathbf{u}$ reaches the solution $\mathbf{Du}_{\mathrm{true}} = \mathbf{b}$. The second term is minimised as R or β become large or as u becomes small. Notice the similarity of this approach and its network with the restoration approach in section 2.2 and its linear NN structure.

Using Kirchoff's current law, we obtain system dynamic equations leading to an error estimation

$$\|\mathbf{u}_{\mathrm{true}} - \mathbf{u}_{\mathrm{final}}\| = \frac{1}{1 + \alpha\beta NR}\|\mathbf{u}_{\mathrm{true}}\| \qquad (5.6)$$

and a time constant estimation

$$\tau = \frac{C}{\alpha\beta N} \qquad (5.7)$$

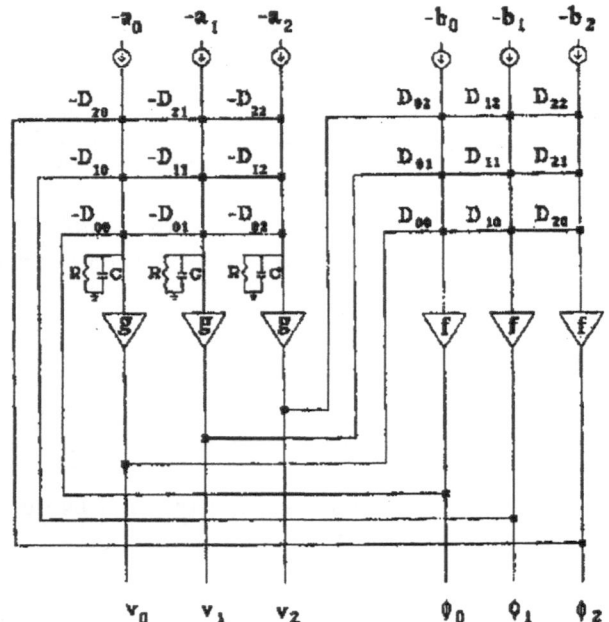

Fig. 5.1. Tank-Hopfield linear programming neural network

6. Conclusions

In this paper, the application of neural networks to multichannel image processing and analysis is presented. Due to their structure, neural networks can be easily implemented in hardware and operate in real time mode. This property renders neural networks a very efficient tool for optimisation problems. It is shown how robust multichannel image restoration can be mapped onto dynamic neural networks, thus taking advantage of great accuracy and high speed of computation.

Moreover, if we assume that the blurring function is completely unknown, the image restoration problem can be viewed as an unknown function interpolation problem, where static neural networks prove to be very efficient, due to their highly non-linear structure and interpolation capabilities.

The optimisation framework is extended to the segmentation of multichannel images, by transforming the cluster assignment problem to that of the minimisation of some measure indicating the dissimilarity of the distribution of pixels in clusters.

Finally, we show how frequently used transformations can be mapped onto a dynamic network by transforming the problem to an equivalent optimisation problem. Obviously, such a rapid transformation approach is extremely interesting, allowing further processing in different domains.

References

1. P. D. Wasserman, *Neural Computing : Theory and Practice*, Van Norstrand Reinhold, New York, 1989.
2. A. Cichoki and R. Unbehauen, *Neural Networks for optimization and signal processing*, Wiley, New York, 1993.
3. J. J. Hopfield, "Neural Networks and physical systems with emergent collective computational abilities", in: *Proceedings National Academy of Sci,ences* vol. 79, pp. 2554–2558, 1982.
4. M. E. Zervakis, "Generalized Maximum A Posteriori Processing of Multichannel Images and Applications", *IEEE Transactions on Circuits, Systems Signal Processing*, vol. 15, no. 2, 1996.
5. N. P. Galatsanos, A. K. Katsaggelos, R. T. Chin and A. D. Hillery, "Least squares restoration of multichannel images", *IEEE Transactions on Signal Processing*, vol. ASSP-39, no. 10, 1991.
6. M. E. Zervakis, "Optimal restoration of multichannel images based on constrained mean-square estimation", *Journal of Visual Communications and Image representation*, vol. 3, no. 4, 1992.
7. A. Katsaggelos, "Iterative image restoration algorithms", *Optical Engineering*, vol. 28, no. 7, 1989.
8. M. Teles de Figueiredo and Jose M. N. Leitao, "Image Restoration Using Neural Networks", *ICASSP Proceedings*, 1992.
9. T. Mu Kwon and M. E. Zervakis, "Design of Regularization Filters Using Linear Neural Networks", *Journal of Artificial Neural Networks*, vol. 1, no. 2, 1994.
10. J. K. Paik and A. K. Katsaggelos, "Image Restoration Using a Modified Hopfield Network", *IEEE Transactions on Image Processing*, vol. 1, no. 1, 1992.
11. Yi-Tong Zhou, R. Chellappa, A. Vaid and B. K. Jenkins, "Image Restoration Using a Neural Network", *IEEE Transactions on Acoustics, Speech and Signal Processing*, vol. 36, no. 7, 1988.
12. K. Sivakumar and U. B. Desai, "Image Restoration Using a Multilayer Perceptron with a Multilevel Sigmoidal Function", *IEEE Transactions on Signal Processing*, vol. 41, no. 5, 1993.
13. B. Kamagar-Parsi, J. A. Gualtieri, J. E. Devany and B. Kamagar-Parsi, "Clustering with neural networks", *Biological Cybernetcs*, vol. 61, pp. 201–208, 1990.
14. L. E. Russo, "An outer product neural net for extracting principal components from a time series", *IEEE Workshop On Neural Networks for Signal Processing*, Princeton, NJ, September 29-October 2, 1991.
15. E. Oja, "Neural networks, principal components and subspaces", *International Journal of Neural Systems*, vol. 1, no. 1, pp. 61–68, 1989.
16. L. E. Russo and E. C. Real, "Image Compression Using an Outer Product Neural Network", *ICASSP Proceedings*, 1992.
17. A. D. Piraino and Y. Takefuji, "Optimization Neural Networks for the Segmentation of Magnetic Resonance Images", *IEEE Transactions on Medical Imaging*, vol. 11, no. 2, 1992.
18. R. C. Gonzalez and R. E. Woods, *Digital Image Processing*, Addison-Wesley Publishing Company, 1992.
19. D. W. Tank and J. J. Hopfield, "Simple 'neural' optimization networks: An A/D converter, signal decision circuit, and a linear programming circuit", *IEEE Transactions Circuits Systems*, vol. CAS–36, pp. 533–541, May 1986.
20. Y. C. Pati, D. Friedman, P. S. Krishnaprasad, C. T. Yao, M. C. Packerar, R. Yang and C. R. K. Marrian, "Neural networks for tactile perception", in *Proceedings 1988 IEEE Conference Robotics and Automation*, Philadelphia, PA.

21. A. D. Culhane, M. C. Peckerar and C. R. K. Marrian, "A Neural Network Approach to Discrete Hartley and Fourier Transforms", *IEEE Transactions on Circuits and Systems*, vol. 36, no. 5, May 1989.

Neural Networks for Classification of Ice Type Concentration from ERS-1 SAR Images

Classical Methods versus Neural Networks

Jan Depenau[1,2]

[1] DAIMI, Computer Science Department,
Aarhus University, Ny Munkegade, Bldg. 540, DK-8000 Aarhus C
e-mail: depenau@daimi.aau.dk
[2] TERMA Elektronik AS, Hovmarken 4, DK-8520 Lystrup
e-mail: depenau@terma.dk

Summary. This paper describes a minor part of the work done in connection with a preliminary investigation of a neural network's capability to classify ice types. It includes a short review of earlier used techniques, implementation of different neural networks and results from various experiments with these networks. The estimation of ice type concentrations from Synthetic Aperture Radar (SAR) images has been investigated for several years, see e.g. [9]. The classification estimation has been performed by training a Bayesian Maximum Likelihood Classifier ($BMLC$) [8] with a classification rate about 80%. The neural networks considered are all of the feed-forward type. For training, different learning algorithms and error functions are used. Both pruning and construction algorithms are used to get an optimal architecture. Experiments showed that almost any kind of neural network, using a Standard Back-Propagation (Std_BP) learning algorithm for minimising the Mean Square Error (MSE), is able to perform better than the $BMLC$. The reason is that the neural network is able to use a larger training set than the $BMLC$.

1. Introduction

A SAR image has fine spatial resolution containing about 8,000 x 8,000 pixels. Before any analysis is carried out the image is segmented. At Electromagnetics Institute (EMI), Technical University of Denmark a program which is able to segment the images has been developed [10]. The segmentation algorithm can be briefly described as: 1) *Edge detection*, 2) *Centre point determination*, 3) *Segment border determination*, and 4) *Segment merging*. For each segment (area) 16 features are calculated. The definition and further explanation of these features can be found both in [5] and [9]. The 16 features are used to classify the ice into one of 6 ice type classes: *Multi-year ice(MY), New ice type a (NIa), New ice type b(NIb), New ice type c (NIc), Open water, (WA)* and *Ice mixture (MIX)*. Each ice type is an indication of how old the ice is. The classification approach that has been used at EMI is the classical *Bayesian Maximum Likelihood Classification* (BMLC) as defined in [8].

Using the segmentation algorithm on a image from the European satellite ERS 1 on a location near the coast of Greenland, the result is an image that is split into more than 20,000 segments. Based on the experience of an expert and general knowledge of the weather conditions and the potential ice types, about half of these segments are classified by visual inspection. Among these segments there are **5,780** which belong to one of the 6 classes of interest distributed in the following way: **3,447** MY, **108** NIa, **693** NIb, **643** NIc, **397** WA **and 492** MIX. Although the number of segments represents only 25% of all segments, the size of the segments constitutes about 50% of all pixels in the image.

The segment's feature is strongly disturbed by noise because the backscatter (BS) coefficient inherits the so-called speckle noise in the SAR image. A statistical analysis of the features made by Skriver in [10] showed among other things that there was a strong correlation among the features of the classes when small areas were considered.

Since the BMLC only works properly with uncorrelated parameters, it is necessary to reduce the variance of the estimated features and the correlations between the features. This is the same as reducing the influence of speckle in the image, which is normally done by simply using as large areas in the image as possible. In [9] only segments with more than 1000 pixels were used in the classification. Combined with the fact that a large number of manually classified segments is in practical applications unrealistic, because it is a very time-consuming and difficult task, only 30 of the largest segments were used for the training set in [9]. This is of course unfortunate because the calculation of the model's parameters only is based on larger segments, but it is to be used for all segments. Contrary to custom these segments were not removed from the test set, but due to the small ratio between the training set and test set this is not considered to be a serious problem. This means that the segments or data were split into a *training set with only* 30 *data* and a *test set with* 5,780 *data*. The training set consists of 5 elements from each class.

1.1 Results from Experiments

Different versions of the model, with a varying number of features, have been calculated and tested. It turned out that the combination of only two features, Mean value of the Back-Scatter coefficient in db (BS) & Power-to-Mean Ratio (PMR) gave the best classification For the two combinations BS and $BS\&PMR$ further results are shown in table 1.1:

Table 1.1. Performance table for different combinations of features measured in percentage of error on the test set.

	BS (in segments)	BS (in area)	BS & PMR (in segments)	BS & PMR (in area)
Error %	43.7	38.1	23.9	27.1

The classification is often used in a context where it is the percentage of the areas that is interesting, because it is more important to classify large areas correctly than small areas. However, in the following investigations the focus will be on the classification ability and therefore the classification will be reported in segments.

The confusion matrix for the "*BS* & *PMR*" classifier is shown in table 1.2:

Table 1.2. Confusion matrix for *BC* & *PMR*. Each row indicates how the elements from the class at the left are classified.

Ice Type	MY	NIa	NIb	NIc	WA	MIX	Total
MY	3145	220	21	6	1	54	3447
NIa	23	69	15	0	0	1	108
NIb	18	156	295	222	2	0	693
NIc	0	0	5	638	0	0	643
WA	178	0	0	0	218	1	397
MIX	434	21	0	0	4	33	492

2. Neural Networks for Classification of Ice

2.1 Neural Network Architecture

In order to explore neural networks' capability to classify ice types, several different networks were implemented. Both construction and pruning algorithms were used to get an optimal architecture. Four types of feed-forward network were considered, a simple perceptron, Multi-layer perceptron, Cascade-Correlation Architecture [2] (CCA) and the Global-Local Architecture [1] (GLOCAL), see figure 2.1.

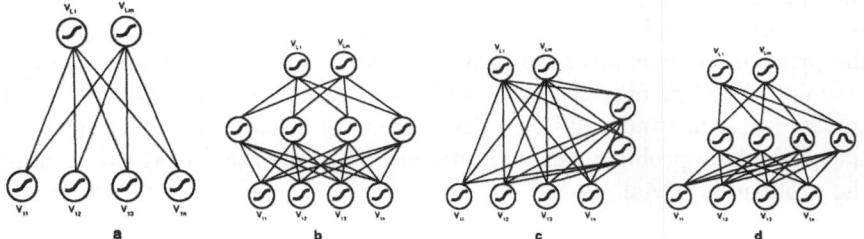

Fig. 2.1. Implemented types of network: a) Simple perceptron, b) Multi-layer perceptron, c) CCA, and d) GLOCAL

Three different pruning algorithms were applied, the Magnitude pruning algorithm [6], Optimal Brain Damages [7] and Optimal Brain Surgeon [4]. For all networks different learning algorithms and error functions are used in the training.

2.2 Experiments and Results on the 30/5780 Data Set

In the first series of experiments with neural networks, the ratio between training and test set was kept at 30/5780 (30 patterns in the training set and 5780 patterns in the test set) in order to be able to compare the results directly with the results from the $BMLC$.

A number of different network architectures have been implemented and trained with different learning algorithms and parameters. The $BMLC$ was able to learn the training set. The requirement to all neural networks, in order to make the result comparable, was therefore that the training should at least continue until the *classification error on the training set was* 0. A representative extract of the networks is shown in table 2.1.

Table 2.1. Performance table for an MLP, the CCA and GLOCAL on the SAR data. *X+Y means X unit in the first hidden layer and Y in the second hidden layer.*

TYPE	Architecture	Number of Weights	Learning Algorithms	Training Set Classification Error in %	Test set Classification Error in %
PERCEP	0	96	Std_BP	0	37.0
PERCEP	0	48	Pru_Mag	0	36.7
MLP	11	242	Std_BP	0	28.7
MLP	7	45	Pru_Mag	0	29.7
MLP	16+3	322	Std_BP	0	32.8
MLP	6+2	56	Pru_Mag	0	36.2
CCA	0	96	Sdt_BP	0	37.0
GLOCAL	0	96	Sdt_BP	0	37.0

The $BMLC$ make a better performance, with 23.9% classification error on the test set, while the best performing network in these experiments has a classification error on the test set of 28.7%. None of these rates are, however, very impressive and the reason is that the training set is too small to represent the problem. This is not surprising for several reasons: the training set was actually picked so only segments with large areas were members, while the smaller segments had a different distribution of features; it is extremely rare that the whole problem can be represented by only 0.5% of the data unless the problem is trivial.

2.3 New Partition of Data

The argument for choosing the small training set was that the $BMLC$ only works properly with a statistical estimate of the features that are central and uncorrelated, and this is only valid for large segments. Neural nets do not take into consideration or make any assumption about the statistical properties of the data, that is one of their strengths[1]. Therefore it makes no sense to

---

[1] Some would argue that this is not the case, see for example [3]

use the data in a better way. There is no problem in selecting all data as training set for a neural network, but since the essential quality is its ability to generalise, i.e. its performance on unknown data, a test set is needed. So the question is how to split the available data in a meaningful and fair way. What is fair then? No strict rules are given, but different opportunities are often used. What is most often seen is that the data set is split up so the training set contains between 30 and 70%, and the test set contains the rest of the elements.

In the following experiments data in the training and test set were chosen at random in such a way that each class in the respective sets contained 50% of all data. No attempt was made to ensure that previous training sets of large segments were included in the new training set. This means that data were split in the following way: a *training set with* 2,892 data, (1,724 *MY*, 54 *NIa*, 347 *NIb*, 322 *NIc*, 199 *WA*, 246 *MIX*) and a *test set with* 2,888 data.

Again a number of different network architectures have been implemented and trained with different learning algorithms and parameters. The requirement to all neural networks, except the third MLP with one hidden layer, was that the training should continue until the MSE did not show any progress. When pruning was performed, the requirement to the pruning was that it should stop as soon as the network made an increase on the MSE that was larger than 10% of the original network's MSE, or when the number of elements misclassified increased by more than 30, and then use the last weight combination that was able to show a classification error on the training set with no more than a 10% increase. In all experiments *shuffle* was used on the training set and it had a vital influence on the network's performance and result. A representative number of the networks designed and implemented is shown in table 2.2.

The best result on the training was performed by the $MLP(16 + 3)$. Table 2.3 shows the training and test confusion matrix for this network.

The perfect confusion matrix would be diagonal with 1724 MY, 54 NIa, 347 NIb, 322 NIc, 199 WA, and 246 MIX and the rest of the matrix entries zero. Note that the network is able to classify some of the NIa members, although the number of training patterns is alarmingly small.

In order to make the results directly comparable with the test confusion matrix for the $BMLC$, the values from table 1.2 should be halved, since the test of the $BMLC$ was performed on a test set with 5780 elements. Comparing table 1.2 and 2.4 clearly reveals why the neural network performs much better than the $BMLC$: it is simply able to learn to classify classes with a relatively large number of patterns. Further investigations will be necessary in order to disclose whether this is due to the quality of a neural network or merely due to the fact that it shows more patterns. For the overall result this might be of minor interest, the important thing is that the network is able to be trained on a large training set, thereby improving the classification rate.

Table 2.2. Performance table for an MLP, the CCA and GLOCAL on the SAR data. *X+Y means X unit in the first hidden layer and Y in the second hidden layer, while X × 1 means X hidden layers with one unit. SCG is an abbreviation for Scaled Conjugate Gradient.*

TYPE	Architecture	Number of Weights	Learning Algorithms	Training Set (Classification Error in %)	Test set (Classification Error in %)
PERCEP	0	96	Std_BP	9.6	11.0
PERCEP	0	50	Pru_OBS	10.4	10.8
MLP	11	242	Std_BP	9.1	11.6
MLP	10	161	Pru_Mag	9.6	12.0
MLP	11	242	Std_BP	10.8	10.8
MLP	16+3	322	Std_BP	8.3	10.7
MLP	16+3	242	Pru_Mag	14.2	14.1
MLP	32+16	1120	SCG	9.6	10.3
CCA	40 × 1	1756	Quick-Prop	12.4	12.8
GLOCAL	46	782	Std_BP	10.0	11.2
PERCEP	0	96	Std_BP	11.0	10.3

Table 2.3. *Training confusion matrix for the MLP(16 + 3) The total number of errors is = 240*

Ice Type	MY	NIa	NIb	NIc	WA	MIX	Errors
MY	1698	21	6	0	2	96	125
NIa	0	6	0	0	0	0	0
NIb	9	23	313	32	0	0	64
NIc	1	0	28	290	0	0	29
WA	0	0	0	0	196	1	1
MIX	16	4	0	0	1	149	21

Another interesting observation is that the type of network, given a data set with such a unequal distribution, is not of significant importance for the performance. So *improvement of the classification rate may be obtained by using a simple perceptron.*

3. Conclusion

Experiments showed that almost any kind of neural network, using a Standard Back-Propagation learning algorithm for minimising the Mean Square error, is able to perform better than the Bayesian Maximum Likelihood Classifier. The reason is that the neural network is able to use a larger training set than the Bayesian Maximum Likelihood Classifier. It turned out that for a given data set with an unequal distribution available for these experiments, a simple perceptron would improve the classification rate compared to the traditional Bayesian Maximum Likelihood Classifier.

Table 2.4. *Test* confusion matrix for $MLP(16 + 3)$. *The total number of errors is* *308 (≈ 10.7%)*

Ice Type	MY	NIa	NIb	NIc	WA	MIX	Errors
MY	1665	17	15	0	15	116	163
NIa	1	0	0	0	0	0	1
NIb	34	28	315	20	0	0	82
NIc	0	1	15	300	0	0	16
WA	2	0	1	1	175	4	8
MIX	22	8	0	0	8	126	38

To do the Bayesian Maximum Likelihood Classifier method justice it should be mentioned that the one that is used at the Electromagnetics Institute, Technical University of Denmark is probably not the optimal one. A larger training set, balanced between the desire of many training data and a growing variance of the estimated features and the correlations between the features in small segments, would probably improve the performance. A better selection of the features used, such as Karhunen-Loeve transformation, would probably improve the result further thereby giving a better foundation for a comparison. However, neural networks could also very well benefit from a Karhunen-Loeve transformation and the networks described in this preliminary investigation are probably also not the best.

The main conclusion is that it would be profitable to apply a neural network in order to classify ice types from the SAR-data. There is, however, no doubt that further research in collaboration with people having expert knowledge of SAR-data is needed, before the results will have a more practical value.

References

1. J. Depenau, "A Global-Local Learning Algorithm", in: *Proceedings from the World Congress on Neural Networks*, vol I, pp. 587–590, Washington 1995.
2. S. E. Fahlman, and C. Lebiere, "The Cascade-Correlation Learning Architecture", in: *Advances in Neural Information Processing Systems II*, (Denver 1989), ed. D. S. Touretzky, pp. 524–532. San Mateo: Morgan Kaufmann, 1990.
3. S. Geman, E. Bienenstock, and R. Doursat, "Neural Networks and the Bias/Variance Dilemma", *Neural Computation* vol. 4, pp. 1–58, 1993.
4. B. Hassibi, and D. Stork, "Second order derivatives for network pruning: Optimal Brain Surgeon", in: *Advances in Neural Information Processing Systems V*, (Denver 1993), ed. S. J. Hanson *et al.*, pp. 164–171. San Mateo: Morgan Kaufmann, 1992.
5. R. M. Haralick, "Statistical and Structural Approaches to Texture", *Proceedings IEEE 67*, pp. 786–804, 1979.
6. J. Hertz, A. Krogh, and R. Palmer, *Introduction to the Theory of Neural Computation*, Addison Wesley, pp. 115–162, 1991.

7. Y. Le Cun, J. S. Denker, and S. A. Solla, "Optimal Brain Damage", in: *Advances in Neural Information Processing Systems II*, (Denver 1989). ed. D. S. Touretzky, pp. 598–605. San Mateo: Morgan Kaufmann, 1990.
8. W. Niblack, *An Introduction to Digital Image Processing*, Strandsberg Publishing Company, Birkerød, Denmark 1985.
9. H. Skriver, "On the accuracy of estimation of ice type concentration from ERS-1 SAR images", from *EARSEL 14th Symposium*, Gothenburg, 1994.
10. H. Skriver, "Extraction of Sea Ice Parameters from Synthetic Aperture Radar Images", Ph.D. Thesis from Electromagnetic Institute, Technical University of Denmark, Lyngby, 1989

A Neural Network Approach to Spectral Mixture Analysis

Theo E. Schouten and Maurice S. klein Gebbinck

Katholieke Universiteit Nijmegen, Computing Science Institute,
Toernooiveld 1, 6525 ED Nijmegen The Netherlands.
e-mail: ths@cs.kun.nl

Summary. Imaging spectrometers acquire images in many narrow spectral bands. Because of the limited spatial resolution, often more than one ground cover category is present in a single pixel. In spectral mixture analysis the fractions of the ground cover categories present in a pixel are determined, assuming a linear mixture model. In this paper neural network methods which are able to perform this analysis are considered. Methods for the construction of training and test data sets for the neural network are given. Using data from 3 spectrometers with 6, 30 and 220 bands and 3 or 4 ground cover categories, it is shown that a back-propagation neural network with one hidden layer is able to learn the relation between the intensities of a pixel and its ground cover fractions. The distributions of the differences between true and calculated fractions show that a neural network performs the same or better than a conventional least squares with covariance matrix method. The calculation of the fractions by a neural network is much faster than by the least squares methods, training of the neural networks requires however a large amount of computer time.

1. Introduction

Imaging spectrometers [1, 2] acquire images simultaneously in many narrow spectral bands from visible to infrared. Their data can be used to determine the physical nature of the materials within the field of view qualitatively and estimate their abundances quantitatively. In ground cover maps each pixel of the acquired image is assigned to only one of the possible ground cover categories. Because of the limited spatial resolution of the scanners used, often more than one category is present in a single pixel (a so-called mixed pixel). In spectral mixture analysis of a pixel the fractions of the ground cover categories present in that pixel are determined. This allows the construction of a mixture map, a series of images showing for each ground cover category its concentration over the area in the image.

In spectral mixture analysis usually a linear mixture model [3] is used, the signal received for a pixel in band i is assumed to be:

$$s_i = \sum_{j=1}^{c} f_j R_{ij} + e_i, \quad i = 1, 2, ..., n \tag{1.1}$$

where n is the total number of bands, R_{ij} is the reflectance of the jth end-member (ground cover category) in the ith band, f_j is the fraction of the pixel covered by end-member j, e_i is the error in the ith band and c is the total number of end-members in the pixel. The purpose of the spectral mixture analysis is to find the best approximation of f knowing the values of s and R and possibly having some information on the statistical properties of e. This analysis gets complicated because of the physical conditions on f: $\sum_j^c f_j = 1$ and $0 \leq f_j$ for all j. Accurate analytical solutions are not available and numerical methods [4, 5] must be used, often iterative ones or methods involving an exhaustive search through the f space.

In this study a back-propagation neural network is used to perform the mapping from s to f. The used images and generation of the data sets used for training and testing of the neural networks are described in section 2. The neural networks used are described in section 3, as well as a comparison of their results with the results of two other methods. In the last section our conclusions are given.

2. Images and Data Set Generation

The first image used was a 30 band image of 512×475 pixels from MAIS[2] (Modular Airborne Imaging Spectrometer) made in China. The resolution of the bands is 20 nm and each pixel corresponds to $20 \times 20m^2$ of the earth surface. Three end-members corresponding to water, vegetation and soil were

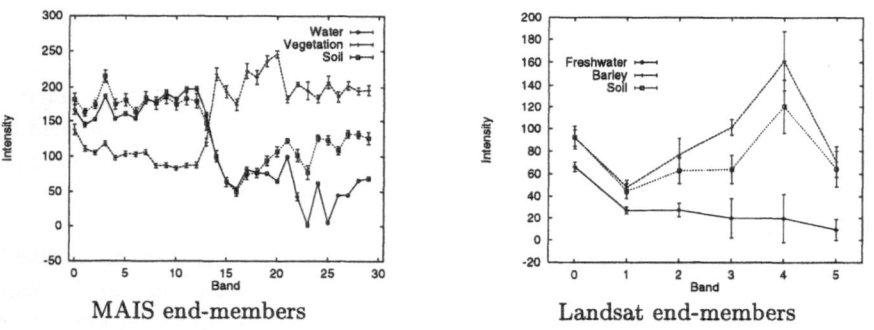

MAIS end-members   Landsat end-members

Fig. 2.1. Average and σ intensities

chosen based on visual inspection and for each of them an example set of 400 pixels was selected. As the second image a 6 band Landsat image (original data Eurimage) of the Lisbon area in Portugal was taken. From the available ground truth data set of 16 ground covers [6] the covers freshwater, barley and soil with 277, 311 and 868 pixels were taken as end-members examples. From the 220 band AVIRIS data accompanying the MultiSpec computer program [7] the covers labelled corn, hay-windrowed and wheat with 234, 245 and 212 pixels were taken as examples for the third end-member set. The bldg-grass-trees-drives cover with 380 pixels was added for the fourth end-member set. The average and σ of the intensities of the end-members are shown in figures 2.1 and 2.2.

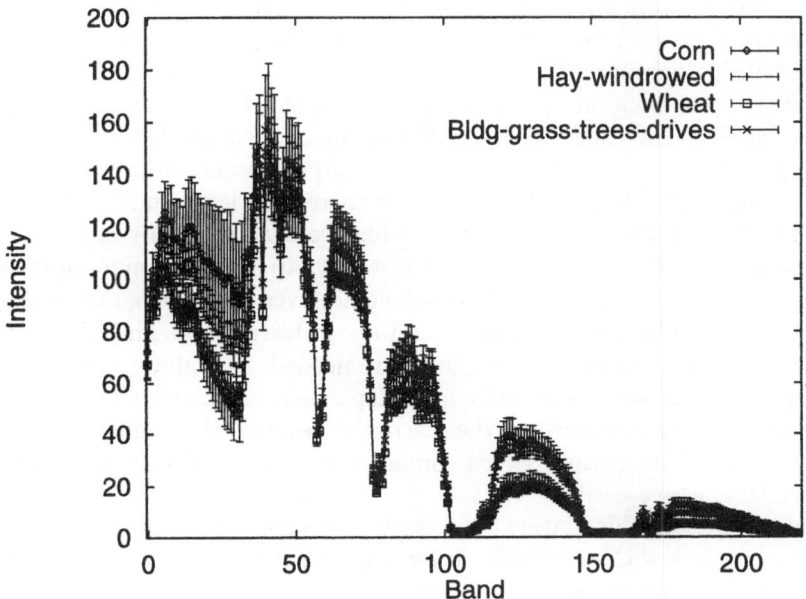

Fig. 2.2. Average and σ intensities of the AVIRIS end-members

For each end-member set 32768 mixed-pixels were generated, half of them are used for training a neural network, the other half for evaluating the accuracy of the f's calculated by the trained neural network. For each mixed-pixel the f's are calculated as:

$$
\begin{aligned}
f_{i\%3} &= r() & f_{i\%4} &= r() \\
f_{i+1\%3} &= r() * (1 - f_{i\%3}) & f_{i+1\%4} &= r() * (1 - f_{i\%4}) \\
f_{i+2\%3} &= 1 - f_{i\%3} - f_{i+1\%3} & f_{i+2\%4} &= r() * (1 - f_{i\%4} - f_{i+1\%4}) \\
& & f_{i+3\%4} &= 1 - f_{i\%4} - f_{i+1\%4} - f_{i+2\%4}
\end{aligned}
$$

where r() is an uniform random number between 0 and 1 and i is increased by 1 for the next pixel. The end-member matrix R_{ij} is then taken as a random selection of pixels from the example sets and the s_i's are calculated as $\sum_{j=1}^{c} f_j R_{ij}$ rounded to an integer in the range of the used spectrometer. The e_i's are thus modelled by the variations in the end-member intensities[5] and the rounding to integer intensity values.

3. Neural Network and Results

As a neural network the multi-layer back-propagation [8] net was used. The simulations were performed with our network simulator program, which executes on a PowerXplorer parallel machine and is implemented in C using

PVM for communication. Based on our experiences with other applications the following strategies were used. The input and output values were scaled linearly to the range -1.0 to 1.0. The transfer functions were taken as *tanh* for the hidden layers and linear for the output layer. The steepness of the transfer function was learned for each neuron[9]. The learning rate was varied by hand during learning, the momentum rate was typically set to 0.8.

Learning was done in epoch mode but weights, biases and steepnesses were updated several times (typically 50) during each epoch. This semi-epoch mode approximates on-line learning which in our experience usually produces better results than full epoch learning when there are a large number of training samples. On a sequential machine on-line learning is usually faster (that is, the weights with best results on the test data are obtained more quickly) than epoch learning. On a parallel machine however different parallelisation strategies[10] can be used for on-line and epoch learning, making the latter faster as less communication operations are needed. The above quality and learning time observations are off course dependent on the type of problem, the structure of its error surface, the size of the network, the number of training patterns and the ratio between communication and calculation speed of the parallel machine used.

The number of hidden layers and neurons appeared to be not very critical. Finally a network with one hidden layer was chosen with 24 neurons for the Landsat and 16 neurons for the other data sets. Because the neural network does not enforce the conditions on f a small post-processing step was added. Negative f values were set to 0 and the f's were then scaled linearly so that their sum equals 1. A very small improvement on the results was observed.

The results of the neural network on the test data are compared to 2 least squares methods, one minimising the function $(s - Rf)^T(s - Rf)$ and the other $(s - Rf)^T C^{-1}(s - Rf)$ (C is the covariance matrix of e) subject to the constraints on f. The minimisations were performed using a 3 level exhaustive search procedure implemented on the parallel machine.

In figures 3.1 to 3.4 the distributions of the difference between the true and determined f_j's are shown. The neural network performs much better than the least squares method. For the MAIS and Landsat data sets the neural network results are a little better than the least squares with C results, for the AVIRIS data sets the results are about the same.

The time taken by a neural network calculation of f can be expressed as the number of (multiply + addition) operations which is approximately given by $h(n + c) + 3h$ with h the number of hidden neurons, assuming the *tanh* function is calculated by table lookup and linear interpolation. For the calculation of the least squares functions nc resp. $n^2 + nc$ operations are needed. The number of steps needed for the exhaustive search procedure is about a constant (around 14 for the used step sizes) to the power $c - 1$. The neural network calculation of f is thus much faster.

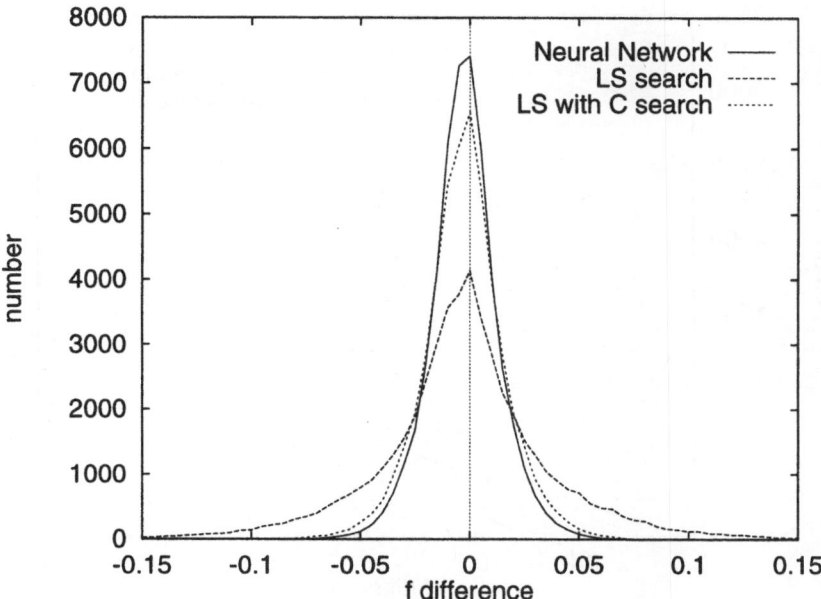

Fig. 3.1. f difference distributions, MAIS 11000 epochs

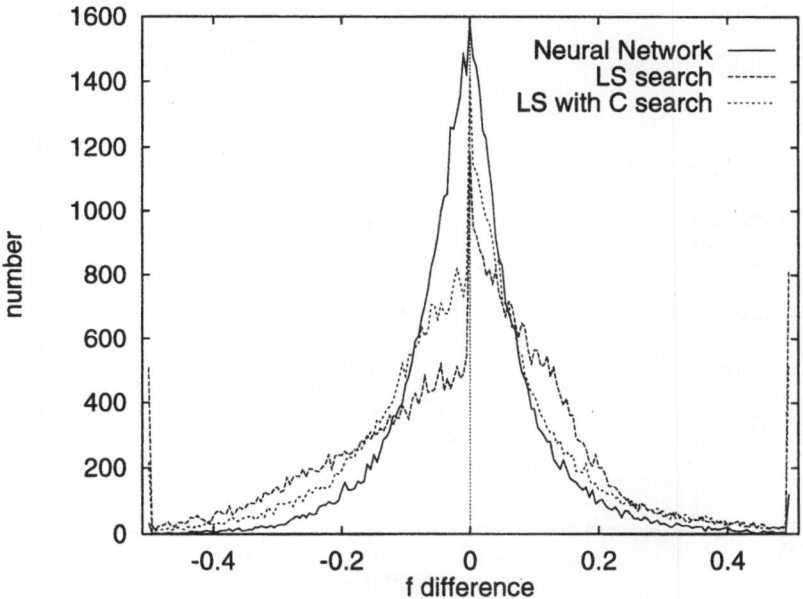

Fig. 3.2. f difference distributions, Landsat 15000 epochs

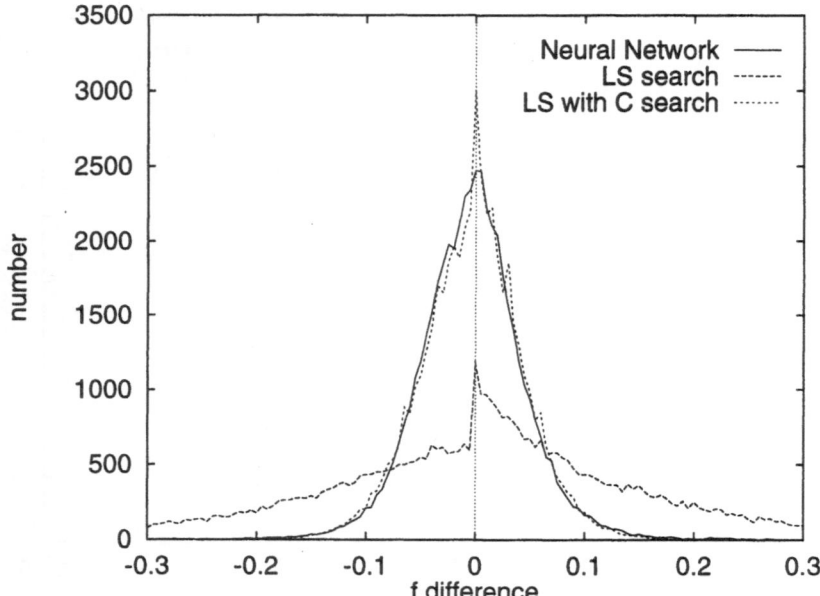

Fig. 3.3. f difference distributions, Aviris-3 17000 epochs

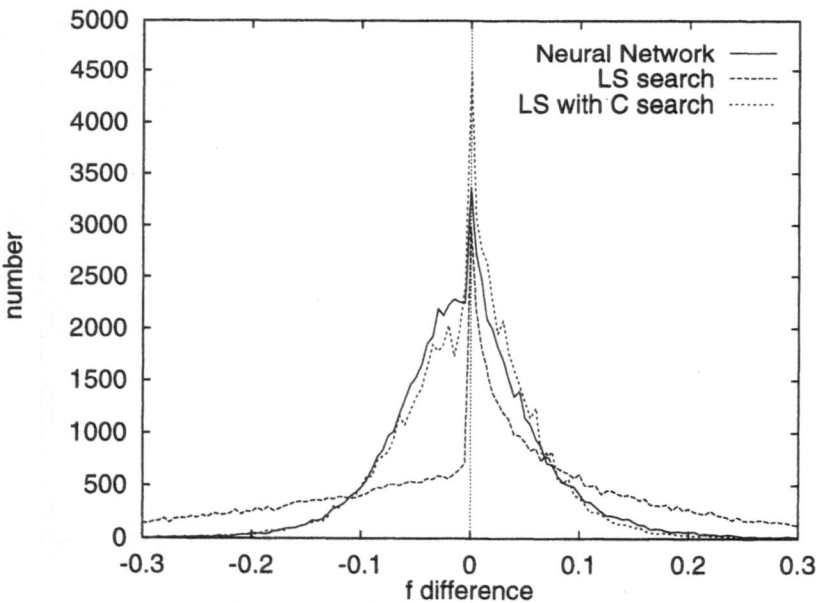

Fig. 3.4. f difference distributions, Aviris-4 34000 epochs

4. Conclusions

A back-propagation neural network with one hidden layer is able to learn the relation between the intensities of a pixel and the fractions of its ground cover categories. This was demonstrated on data from 3 spectrometers with 6, 30 and 220 bands using 3 end-members and also 4 for the 220 band data. The distribution of the difference between true and calculated fractions shows that a neural network performs the same or better than a conventional least squares with covariance method and clearly better than a simple least squares method.

Training of the neural networks requires the construction of training and test data sets and is relatively straightforward but very time consuming. The calculation of the fractions by a neural network is however much faster than calculation by the least squares methods.

References

1. G. Vane et al, "The Airborne Visible/Infrared Imaging Spectrometer(AVIRIS)", *Remote Sensing of Environment*, vol. 44, pp. 127–143, 1993.
2. Y. Q. Xue, "The Technical Development of Airborne Scan Imaging Systems", *Journal of Infrared and Millimetre Waves (China)*, vol. 11, pp. 168–191, 1992.
3. H. M. Horwitz, R. F. Napelka, P. D. Hyde and J. P. Morgenstern, "Estimating the proportion of objects within a single resolution element of a multispectral scanner", *Proceedings of the 7th International Symposium on Remote Sensing of Environment*, pp. 1307–1320, 1971, Ann Harbor, Michigan.
4. J. J. Settle, and N. A. Drake, "Linear mixing and the estimation of ground cover proportions", *International Journal of Remote Sensing*, vol. 14, pp. 1159–1177, 1993.
5. M. S. klein Gebbinck, and T. E. Schouten, "Decomposition of mixed pixels", *Image and Signal Processing for Remote Sensing II*, Jacky Desachy, ed., Proceedings SPIE 2579, 1995.
6. R. P. H. M. Schoenmakers, G. G. Wilkinson, and T. E. Schouten, "Results of a hybrid segmentation method", *Image and Signal Processing for Remote Sensing I*, Jacky Desachy, ed., Proceedings SPIE 2315: pp. 113–126, 1994.
7. D. Landgrebe, "Multispectral Data Analysis: A Signal Theory Perspective", Purdue University June 1996,
   http://dynamo.ecn.purdue.edu/~biehl/MultiSpec/
8. D. E. Rumelhart, G. E. Hinton, and R. J. Williams (Eds.), "Learning internal representations by error propagation", in: *Parallel Distributed Processing*, (chap. 8, pp. 318–362), Cambridge MA, MIT Press 1986.
9. A. Sperduti, and A. Starita, "Speed up learning and network optimization with extended back propagation", *Neural Networks*, vol. 6, pp. 365–383, 1993.
10. L. Vuurpijl, T. Schouten, and J. Vytopil, "Performance Prediction of Large MIMD Systems for Parallel Neural Network Simulations", *Future Generation Computer Systems*, vol. 11, pp. 221–232, 1995.

Comparison Between Systems of Image Interpretation

Jean-Pierre Cocquerez, Sylvie Philipp, and Philippe Gaussier

ETIS, ENSEA, 6 avenue du Ponceau, 95014 Cergy-Pontoise Cedex, FRANCE
e-mail: philipp@ensea.fr

Summary. In this paper, we present classification methods, which have been adapted to the pattern recognition problem in the domain of aerial imagery. Three methods based on different mechanisms have been developed: rule-based system, neural networks and fuzzy classification. Our contributions are the adaptation of these methods to the concrete case of aerial images, and the quantitative comparison between the three types of methods.

1. Introduction

The analysis of aerial images presents difficulties because they are composed of a large number of various objects which may take different forms or be located in different relative situations.

In this paper we present classification methods which have been adapted to the pattern recognition problem, in the domain of aerial imagery. Three methods based on different mechanisms have been developed: rule-based system, neural networks and fuzzy classification. Our contributions are the adaptation of these methods to the concrete case of aerial images, and the quantitative comparison between the three types of methods.

Prior to the interpretation, we process a segmentation [1] by contour extraction and closing, and connected component extraction. The image is then represented by an adjacency graph of regions. A classification from the features associated to each node of the adjacency graph is performed, instead of using specialist operators [2], or matching the image description with a model [3, 4]. The system of interpretation is composed of two parts. The first one is an *initial labelling* which gives a classification of each region depending only on the features of the region, regardless of the neighbourhood. The second part aims at giving a coherent labelling to the whole image; this is gradually achieved by a *constraint propagation system*. Both parts are performed by rule-based systems, neural networks and a fuzzy classifier.

The *rule-based system* models our knowledge of the scene; it requires no learning. The initial labelling separates the various classes by hyperplanes, which is the simplest way to manage the classification but which poses the problem of the choice of the thresholds. Another difficulty arises in generalisation, an important task with images of various types for which new rules are necessary.

The *neural implementation* solves the problem of the rule development, as well as the choice of the thresholds, since they are automatically determined by the system during the learning phase (Gradient Back Propagation) [5]. The main difficulties here are the dimensionality of the neural networks and the "multiple credit assignment problem". We will give solutions for both of these problems.

The *fuzzy classification* does not make any assumption about the class shapes, which are dynamically modified during the classification. A learning set is required to initialise the classes. It provides for each region a confidence measure for belonging to each class.

2. Initial Labelling

For each region **R**, the classification is performed from the components of a feature vector **A** of 4 components: surface, average grey level, approximation by a parallelogram (shape factor), and compactness (or circularity). Each of the three implementations of this initial labelling supplies a set of possible labels (6 classes: lawn, grove, building, tree, shadow and vertical face).

2.1 Initial Labelling by a Rule-Based System

To each feature are associated several predicates which take the value true or false. The predicates use thresholds which have been fixed empirically after examination of some simple statistics elaborated on several images. The expert system assigns labels to a region **R** using rules such as:

IF condition(**R**) THEN label(**R**)

The term condition(**R**) is a disjunction of conjunctions of the above predicates. For example, a region **R** is labelled shadow by the following rule:

IF (very_dark(**R**) AND small(**R**) AND compact(**R**))
OR (very_dark(**R**) AND NOT(small(**R**)) AND NOT(compact(**R**)))
THEN label(**R**,shadow)

The rules are not numerous (about twenty); the thresholds values are not critical and allow a multiple labelling of the regions. The rule-based classifier used is a very classical one, but it provides useful information with which to chose the architecture of the neural networks.

2.2 Initial Labelling by a Neural Network

A multi-layer perceptron is well suited to achieve the initial labelling; the learning is performed by the Gradient Back Propagation algorithm [7]. A solution to sizing the neural network can be proposed here: the one we used is composed of four layers, with two hidden layers. The *input layer* is made of 4 neurons, each of them being associated to a region feature. The *output layer* contains as many neurons as classes (6).

The *first hidden* layer acts by thresholding the input values, it receives the input vector in the following form:

$$\mathbf{E} = [\mathbf{e_1}, \ldots, \mathbf{e_i}, \ldots, \mathbf{e_N}]^t$$

For a given region **R**, the components e_i are the normalised analog components a_i of the vector $\mathbf{A} = [a_1, \ldots, a_i, \ldots, a_M]^t$. In the rule based classifier, a maximum of 4 predicates are defined from each analog input (for instance: dark(**R**). So, each neuron of the input layer is connected to 4 neurons of the first hidden layer. The chosen structure limits the connections between the first two layers, which simplifies the problem of thresholding, and insures a better control of the neuron behaviour. Independent learning for the threshold of each input is guaranteed and the ability for the net to generalise is preserved (it is prevented from learning a particular configuration of the input set). The structure is quite regular: all neurons of the first hidden layer receive the same number of inputs whatever the analog variable they process.

The *second hidden layer* is sized in accordance with the rule-based system complexity. A rule may be emulated by a neural network. For instance, figure 2.1 shows how one neuron may realize a logical AND. It is also easy to simulate a logical OR. As a rule in the expert-system is a disjunction of conjunctions, we chose to perform an AND operator with each neuron of the second hidden layer and an OR with each neuron of the output layer. The size of the hidden layers are deduced from the structure of the rules of the expert-system: these are disjunctive rules which use at the most 3 conjunctions of predicates. So, 3 neurons of the second hidden layer are connected to each output neuron and consequently, the second hidden layer contains 18 neurons.

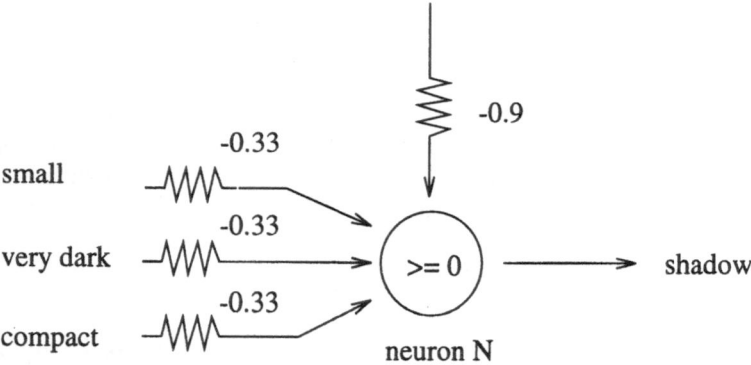

Fig. 2.1. Neural network structure for the rule: IF the region is (small, very_dark, compact) THEN the region may be a shadow

The global architecture is then a 4-16-18-6 network, which is represented in figure 2.2.

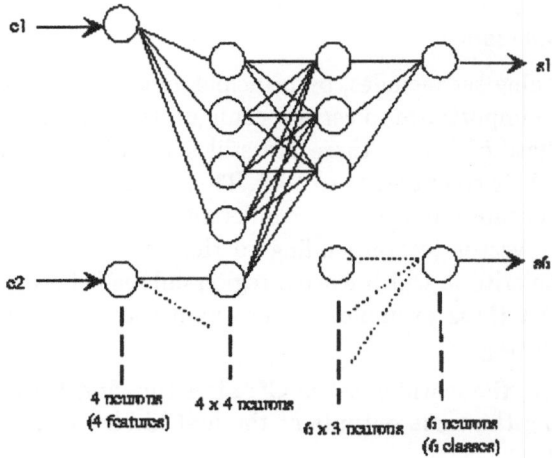

c1

c2

**4 neurons
(4 features)**   **4 x 4 neurons**   **6 x 3 neurons**   **6 neurons
(6 classes)**

s1

s6

Fig. 2.2. Neural network structure

Let us now explain the initialisation. The learning set is composed of examples:

$$C^e = \{(E_1, S_1), \ldots, (E_h, S_h), \ldots, (E_L, S_L)\}$$

For a region R_h, E_h is the vector A_h normalised and S_h is the output vector. If S_h corresponds to a region labelled by the expert, it owns a single "+1" component, if not, all its components equal "-1".

The problem we met in the learning phase was the multiple credit assignment problem:

- two regions with different labels may have almost the same input vector (leading to the same output vector on the first hidden layer).
- a non labelled region may have almost the same input vector as a labelled region.

These phenomena lead to a non-convergence of the network or to a convergence towards a non significant state: the weights are not properly computed because of the contradictory examples, and the system may not even label a region which had a label in the learning set! To solve this problem, we modified the learning set as follows:

- when two regions R_h and R_p produce two examples (E_h, S_h) and (E_p, S_p) with the same binary vector $(H_h = H_p)$ as output of the first hidden layer, both examples are kept but the output vectors S_h and S_p are replaced by a new vector:
$S_h^f = S_p^f = S_h \vee S_p$ (f means "fusion" of the vectors) (1)
obtained by achieving a logical OR for every component of the output vector (-1 is treated as a logical 0 for this operation).

– this process is applied iteratively to all regions, leading to a new learning set with multiple labels.

The new learning set includes, by principle, identical examples. This is a way to give more importance to certain configurations during the learning.

After this phase of fusion, there are still some examples whose output vectors have all their components to "-1". These examples are suppressed in the learning set, because they may have different input vectors, despite having the same output vector (corresponding to the "unknown" class). Another reason is that the criteria which leave a region unlabelled are not always the same, but at least these examples are not numerous. The learning is thus performed in two stages:

– in the first step, the learning set is $\mathbf{C^e}$; after this first learning stage, the examples giving the same outputs at the first hidden layer are merged as in relation (1),
– in the second step, the learning is performed with a learning set which contains multiple labels:

$$\mathbf{C^{fe}} = \left\{ \left(\mathbf{E_1}, \mathbf{S_1^f}\right), \ldots, \left(\mathbf{E_h}, \mathbf{S_h^f}\right), \ldots, \left(\mathbf{E_K}, \mathbf{S_K^f}\right) \right\}$$

2.3 Initial Labelling by Fuzzy Classification

The third method uses a non parametric and non optimal algorithm which works on the shape of the classes in the feature space and modifies this shape as new regions are classified [6]. It provides for each region a degree of membership to classes ([0,1]), instead of binary labels as in the two previous approaches.

The membership grade is given by a potential function which represents a measure of the proximity of the region to the class. It depends on the local density of data previously labelled (actually the k nearest neighbours).

The learning phase aims at building the potential functions for all classes and determining the threshold of membership to the classes (which is the same for all classes).

The algorithm allows an overlapping of the classes, which corresponds to a multiple labelling.

2.4 Results

The learning for the rule-based and the connectionist methods was performed with a learning set of about 600 regions belonging to two images. The classification of these images gives a percentage of good classified regions of 62 for the rule-based system, 30 for the neural network and of 67 for the fuzzy classifier (only the regions with a single label enter in the accounts). For a test-image, where no region was learnt (figure 2.3), the results are: 62% of good classified regions for the rule-based system, 41% for the neural network and 58% for the fuzzy classifier.

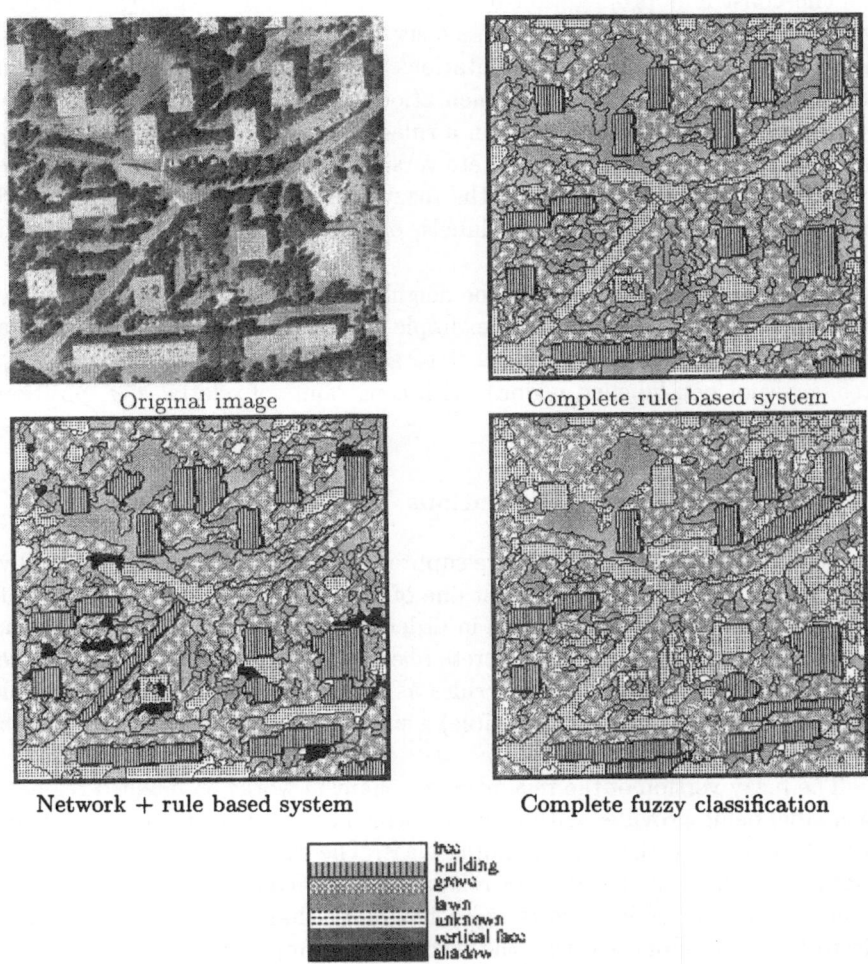

Fig. 2.3. Interpretation of an image different from the learning set

3. Final Labelling by Constraint Propagation

At the end of the initial labelling, achieved by one of the three methods, each region has a set of possible labels and weights. The problem is to refine this labelling and to attach a single label to each region, consistent with the labels of the adjacent regions. This is the role of the constraint propagation system.

The constraint propagation by relaxation was introduced by Rosenfeld who described in [8] three versions, a discrete one, a fuzzy one and a stochastic one. We made three implementations of the constraint propagation, in accordance with our three implementations of the initial labelling: an improvement of the discrete version in a rule-based system, an original connectionist implementation of the discrete version and the fuzzy version after the fuzzy classifier. The discrete and the fuzzy versions, proposed by Rosenfeld only decrease the weights of the labels, or suppress labels, they never add labels nor increase their weights.

The three methods use the same neighbourhood constraints, dictated by our knowledge of the scene. For example a tree may not be included in a building, a shadow must be adjacent to a structure high enough (building, tree, grove or non-labelled region), with constraints of relative size, position and boarder length.

3.1 Discrete and Fuzzy Relaxations

In the rule-based system, labels are suppressed step by step; a label is finally kept if it is consistent with at least one of its neighbours. Our improvement is to add "obligation" constraints, in order to force the labelling of a region: after suppressing labels by the discrete relaxation, regions without a label are examined, new rules or the same rules as before but with looser thresholds are used. A region receives (if possible) a single label, which is consistent with the surrounding regions.

The fuzzy version of the relaxation mechanism works on labelled regions, each label being provided with a weight, which corresponds to the confidence of its belonging to the corresponding class. The compatibility rules are deduced from those of the discrete relaxation: each constraint is modelled by a fuzzy function, which allows more flexibility than the thresholds of the discrete version. For each rule the degree of incompatibility is obtained by using the fuzzy operator minimum on each constraint, thus carrying out a conjunction of facts.

3.2 A Neural Network for the Constraint Propagation

The constraint propagation uses the features of the region, the features of the adjacent regions and the features of the adjacencies. One of the problems is to size the input vector because the number of adjacent regions is variable.

For a region $\mathbf{R_h}$, the input vector $\mathbf{P_h}$ is composed of three parts (cf. figure 3.1):

- the labels of the current region, $\mathbf{S_h}$
- the labels of the adjacent regions after merging by a logical OR $\bigvee_j \mathbf{S_j}$, as exposed above.
- the features of the adjacency constraints, also merged by a logical OR $\bigvee_j \mathbf{F_{hj}}$

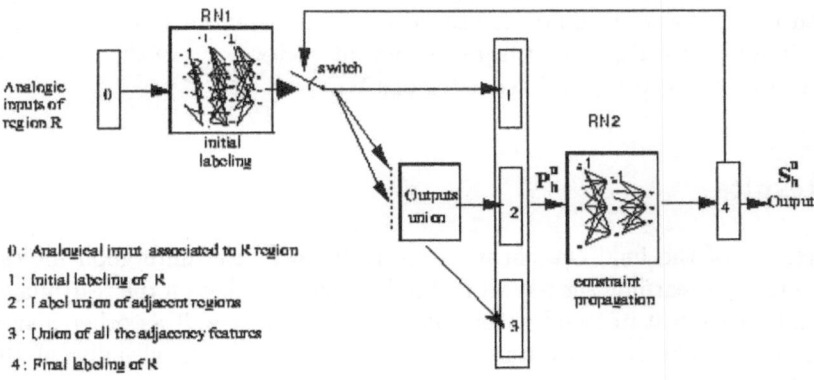

Fig. 3.1. A system of neuron networks for image interpretation

The binary vector $\mathbf{F_{hj}}$ characterises the compatibility between the labels of the region $\mathbf{R_h}$, and those of the region $\mathbf{R_j}$. It is a vector of 8 components, which expresses the consistency of each label of $\mathbf{R_h}$ with those of $\mathbf{R_j}$. The same constraints as for the rule-based system are used.

The logical unions of the labels of the adjacent regions and of the adjacency constraints are sufficient because a label has to be compatible with at least one of the labels of one of the adjacent regions to be kept.

Besides, the network must be able to manage with multiple labels, single label and no label. So it must learn, on one hand, examples $(\mathbf{P_h}, \mathbf{S_h})$ issued from the learning set $\mathbf{C^e}$ which include at most one label per region, and on the other hand, examples $(\mathbf{P'_h}, \mathbf{S_h})$ issued from the learning set of the initial labelling $\mathbf{C^{fe}} = \{(\mathbf{E_1}, \mathbf{S_1^f}), \ldots, (\mathbf{E_h}, \mathbf{S_h^f}), \ldots, (\mathbf{E_K}, \mathbf{S_K^f})\}$ and which may include several labels. The learning set is then:

$$\mathbf{C^P} = (\mathbf{P_1}, \mathbf{S_1}), \ldots, (\mathbf{P_h}, \mathbf{S_h}), \ldots, (\mathbf{P_K}, \mathbf{S_K}), \ldots,$$
$$(\mathbf{P'_1}, \mathbf{S_1}), \ldots, (\mathbf{P'_h}, \mathbf{S_h}), \ldots, (\mathbf{P'_L}, \mathbf{S_L})$$

The structure of the neural network was empirically determined, because it is difficult to deduce the dimension criteria from the rule-based system. It includes an input layer with 20 (6+6+8) neurons and an output layer of 6 neurons (for the 6 classes). Various attempts with zero and one hidden layer were performed. The best results were obtained with a hidden layer of 21 neurons.

During the classification of the regions, this second neural network works in two phases, corresponding to the two kinds of examples:

In the first phase, the neural net receives as input vectors, the vectors built $\mathbf{P}_h^1$ from the results of the initial labelling. The output vector corresponding to $\mathbf{P}_h^1$ is named $\mathbf{S}_h^1$.

In the second phase, the labelling must be propagated through the image. The output of the net is re-injected towards the input (figure 3.1). At iteration n, the input vector $\mathbf{P}_h^n$ is built from the output vector $\mathbf{S}_h^{n-1}$ of the region $\mathbf{R}_h$ and the output vectors of adjacent regions $\mathbf{R}_j$.

4. Results

The results of the final classification see (table 4.1) are quite good for the images used as learning set but all methods make mistakes on the test-image, although other features and other rules were tried, as well as other neural architectures. The results obtained are quantitatively and visually, of the same order for the three methods.

The discrete and the fuzzy relaxations efficiently solve the problem of multiple labelling when the initial labelling is good enough. While the relaxation algorithm keeps the labels lawn, grove, building and tree, it tends to reduce the label shadow and vertical face which are more constrained.

Table 4.1. Numerical results of interpretation of image of figure 2.3

classes	number of regions	neural network and rule-based system		Rule Based System		fuzzy classification	
		neural network	final labelling	initial labelling	final labelling	initial labelling	final labelling
building	16	13 (81%)	13 (81%)	12 (75%)	12 (75%)	9(66%)	10 (62%)
shadows	7	0	0	1 (14%)	0	0	0
vertical face	0	0	0	0	0	0	0
tree	10	0	0	0	0	1(10%)	1(10%)
grove	62	38(61%)	41(66%)	45(73%)	45(73%)	46(74%)	45(73%)
lawn	67	16(24%)	44(66%)	43(64%)	48(72%)	38(57%)	35(52%)
TOTAL number of regions	162	67(41%)	98(60%)	101(62%)	105(65%)	94(58%)	91(56%)
percentage of well classified pixels			77%		82%	0	0

The various simulations have shown that the performance of the second neural network is not very dependent upon the results of the initial labelling (it compensates for the errors made by the first one using the information relating to the adjacent regions). It is actually not interesting to oblige the first network (RN1) to solve a great number of particular cases, because, on the one hand, it would need a more complex network and on the other hand, one would take the risk of learning only particular cases which would produce aberrant results on types of regions never encountered. When the ability to generalise with new examples is needed or when the test phase has to be quick, neural networks are a good solution.

As rule-based and connectionist modules are interchangeable, a good solution is to use a "mixed" method which combines the initial labelling by the connectionist method and the constraint propagation by the rule-based system (figure 2.3). The symbolic approach in the rule-based system is limited by the simplicity of the rules used in the initial labelling and by the choice of the thresholds, while it is an efficient way to model neighbourhood constraints. The neural network for constraint propagation is difficult to size and a modification of the neighbourhood relations implies resizing it. Practically, we do not use this network.

The fuzzy classification may be used when the clusters are ill defined, but its main advantage is that the results it provides may be used by latter processes [9].

The most important fact is that these methods are complementary. The rule-based system allows symbolic knowledge representation. Its study gives useful information to structure and to size the neural classifier. Besides, the latter could be transformed into a fuzzy classifier, by replacing the output function of the last layer by a radial basis function. Some problems remain; for instance, the quality of the results is strongly tied to the quality of the segmentation. In recent works, we proposed a solution based on the co-operation of several segmentation operators [9]. Besides, the number of examples in our learning set is still too low. At present we are building a more representative basis of aerial pictures within the framework of a co-operation with the French National Institute of Geography.

References

1. J-P. Cocquerez, and S. Philipp, "Analyse d'images : Filtrage et segmentation", *ed Masson*, Paris, 1995.
2. M. Nagao, and T. Matsuyama, "Edge preserving smoothing", *Computer vision, graphics, and image processing*, vol. 9, pp. 394–407, 1979.
3. A. Huertas, and R. Nevatia, "Detecting buildings in aerial images", *CVGIP*, vol. 41, pp. 131–152, 1988.

4. D. Mc Keown, W. A. Harvey, and J. Mc Dermott, "Rule-based interpretation of aerial imagery", *IEEE Transactions on Pattern Analysis and Machine Intelligence*, vol. 7, no. 5, 1985.
5. D. E. Rumelhart, G. E. Hinton, and R. J. Williams, "Learning internal representations by error propagation", in: *Parallel Distributed Processing: Explorations in the Microstructures of Cognition*, vol. 1, Cambridge, MA, MIT Press, 1986.
6. B. Charroux, and S. Philipp, "Interpretation of aerial images based on potential functions", *9th Scandinavian Conference on Image Analysis*, Uppsala, june 1995.
7. J-P. Cocquerez, P. GAUSSIER, S. Philipp, and M. Zahid, "Expert-system and connectionist networks for complex aerial picture analysis", *7th Scandinavian Conference on Image Analysis*, Aalborg, August 1991.
8. A. Rosenfeld, R. A. Hummel, and S. W. Zucker, "Scene labelling by relaxation operations", *IEEE Transactions on System, Man and Cybernetics*, vol. SMC-6, pp. 420-433,1976.
9. B. Charroux, S. Philipp, and J-P. Cocquerez, "Image analysis: segmentation operator cooperation led by the interpretation", *IEEE International. Conference on Image Processing*, Lausanne, 1996.

Feature Extraction for Neural Network Classifiers*

Jon A. Benediktsson and Johannes R. Sveinsson

Engineering Research Institute, University of Iceland
Hjardarhaga 2–6, 107 Reykjavik, Iceland

Summary. Classification of multi-source remote sensing and geographic data by neural networks is discussed with respect to feature extraction. Several feature extraction methods are reviewed, including principal component analysis, discriminant analysis, and the recently proposed decision boundary feature extraction method. The feature extraction methods are then applied in experiments in conjunction with classification by multilayer neural networks. The decision boundary feature extraction method shows excellent performance in the experiments.

1. Introduction

Representation of input data for neural networks is important and can significantly affect the classification performance of neural networks. The selection of input representation is related to the general pattern recognition process of selecting input classification variables which strongly affect classifier design. This means if the input variables show significant differences from one class to another, the classifier can be designed more easily with better performance. Therefore, the selection of variables is a key problem in pattern recognition and is termed feature selection or feature extraction [1]. Feature extraction can, thus, be used to transform the input data and in some way find the best input representation for neural networks.

For high-dimensional data, large neural networks (with many inputs and a large number of hidden neurons) are often used. The training time of a large neural network can be very long. Also, the training methods for neural networks are based on estimating the weights and biases for the networks. If the neural networks are large, then many parameters need to be estimated based on a finite number of training samples. In that case, overfitting can possibly be observed, that is, the neural networks may not generalise well although high classification accuracy can be achieved for training data. Also, for high-dimensional data the curse of dimensionality or the Hughes phenomenon [1]

* The authors are very grateful to Dr. Chulhee Lee of the National Institute of Health (NIH) and Prof. David A. Landgrebe of Purdue University for their invaluable assistance in preparing this paper. The authors also thank Dr. Joseph Hoffbeck of AT&T for providing his Matlab data analysis software to us. The Anderson River SAR/MSS data set was acquired, preprocessed, and loaned by the Canada Centre for Remote Sensing, Department of Energy Mines, and Resources, of the Government of Canada. This work was funded in part by the Icelandic Science Council and the Research Fund of the University of Iceland.

may occur. Hence, it is necessary to reduce the input dimensionality for the neural network in order to obtain a smaller network which performs well both in terms of training and test classification accuracies. This leads to the importance of feature extraction for neural networks, that is, to find the best representation of input data in lower dimensional space where the representation does not lead to a significant decrease in overall classification accuracy as compared to the one obtained in the original feature space. However, few feature extraction algorithms are available for neural networks. Though some of the conventional feature extraction methods, such as principal component analysis (PCA) and discriminant analysis (DA), might be used, such methods do not take full advantage of the way neural networks define complex decision boundaries.

Several authors have proposed the use of neural networks for feature extraction [2, 3, 4]. All these authors concentrate on proposing neural networks which do feature extraction. In their works, the neural networks can be nonlinear and either supervised or unsupervised feature extractors. However, they did not focus on data representation and feature extraction for neural networks. In contrast, a neural network for nearest neighbour classification and linear feature extraction is proposed in [5], but their feature extractor is not specified and they do no analysis of different feature extraction methods.

Of interest here is to investigate what kind of linear feature extraction is desirable for neural networks. Linear feature extraction of input data for neural networks should be beneficial since simpler classifiers can be trained more easily with low dimensional data than high dimensional data.

This paper applies several feature extraction methods for neural networks to reduce multi-source remote sensing and geographic data to relatively few features. The goal is to do this without much loss in overall classification accuracy. Feature extraction methods will be discussed in section 2 and experimental results are given in section 3.

2. Feature Extraction

Feature extraction can be viewed as finding a set of vectors that represent an observation while reducing the dimensionality. In pattern recognition, it is desirable to extract features that are focused on discriminating between classes. Although a reduction in dimensionality is desirable, the error increment due to the reduction in dimension has to be without sacrificing the discriminative power of classifiers. The development of feature extraction methods has been one of the most important problems in the field of pattern analysis and has been studied extensively. Feature extraction methods can be both unsupervised and supervised, and also linear and nonlinear. Here we concentrate on linear feature extraction methods for neural networks and then leave the neural networks with the classification task.

The question of how input data should be represented for a neural network is an important one and strongly affects the classification performance of the neural network. Some authors [6, 7], have suggested PCA as a feature extraction method for neural networks but here it will be shown that the method is not optimal in terms of classification accuracy. Below three different linear feature extraction methods are discussed. For all these methods a feature matrix is defined and the eigenvalues of the feature matrix are ordered in decreasing order along with their corresponding eigenvectors. The number of input dimensions corresponds to the number of eigenvectors selected [1]. The transformed data are determined by

$$Y = \Phi^T X \tag{2.1}$$

where Φ is the transformation matrix composed of the eigenvectors of the feature matrix, X is the data in the original feature space, and Y is the transformed data in the new feature space.

2.1 Principal Component Analysis

One of the most widely used transforms for signal representation and data compression is the principal component (Karhunen-Loeve) transformation.

To find the necessary transformation for X to Y in (2.1), the global covariance matrix for the original data set Σ_X is estimated. Then the eigenvalue-eigenvector decomposition for the covariance matrix Σ_X is determined, that is,

$$\Sigma_X = \Phi \Lambda \Phi^T \tag{2.2}$$

where Λ is a diagonal matrix with the eigenvalues of Σ_X in decreasing order and Φ^T is a normalised matrix with corresponding eigenvectors of Σ_X. With this choice of the transformation matrix in (2.1), it is easily seen that the covariance matrix for the transformed data is $\Sigma_Y = \Lambda$.

Although, the principal component transformation is optimal for signal representation in the sense that it provides the smallest mean squared error for a given number of features, quite often the features defined by this transformation are not optimal with regard to class separability [1]. In feature extraction for classification, it is not the mean squared error but the classification accuracy that must be considered as the primary criterion for feature extraction.

2.2 Discriminant Analysis

The principal component transformation is based upon the global covariance matrix. Therefore, it is explicitly not sensitive to inter-class structure. It often works as a feature reduction tool because classes are frequently distributed in the direction of the maximum data scatter. Discriminant analysis is a method which is intended to enhance separability. A within-class scatter matrix, Σ_W, and a between-class scatter matrix, Σ_B, are defined:

$$\Sigma_W = \sum_i P(\omega_i)\Sigma_i \tag{2.3}$$

$$\Sigma_B = \sum_i P(\omega_i)(M_i - M_0)(M_i - M_0)^T \tag{2.4}$$

$$M_0 = \sum_i P(\omega_i)M_i \tag{2.5}$$

where M_i is the mean vector for the $i-th$ class, Σ_i is the covariance matrix for the $i-th$ class, and $P(\omega_i)$ is the prior probability of the $i-th$ class. The criterion for optimisation may be defined as

$$J = \text{tr}(\Sigma_W^{-1}\Sigma_B) \tag{2.6}$$

where tr( ) denotes the trace of a matrix. New feature vectors are selected to maximise the criterion.

The necessary transformation from X to Y in (2.1) is found by taking the eigenvalue-eigenvector decomposition of the matrix $\Sigma_W^{-1}\Sigma_B$ and then taking the transformation matrix as the normalised eigenvectors corresponding to the eigenvalues in decreasing order. However, this method does have some shortcomings. For example, since discriminant analysis mainly utilises class mean differences, the feature vectors selected by discriminant analysis are not reliable if mean vectors are near to one another. Since the lumped covariance matrix is used in the criterion, discriminant analysis may lose information contained in class covariance differences. Also, the maximum rank of Σ_B is $M-1$ since Σ_B is dependent on M_0. Usually Σ_W is of full rank and, therefore, the maximum rank of $\Sigma_W^{-1}\Sigma_B$ is $M-1$. This indicates that at maximum $M-1$ features can be extracted by this approach. Another problem is that the criterion function in (2.6) generally does not have direct relationship to the error probability.

2.3 Decision Boundary Feature Extraction (DBFE)

Lee and Landgrebe [8] showed that discriminantly informative features and discriminantly redundant features can be extracted from the decision boundary itself. They also showed that discriminantly informative feature vectors have a component which is normal to the decision boundary at least at one point on the decision boundary. Further, discriminantly redundant feature vectors are orthogonal to a vector normal to the decision boundary at every point on the decision boundary.

Lee and Landgrebe [8] use a non-parametric procedure to find the decision boundary numerically. From the decision boundary, normal vectors, $N(X)$, are estimated using a gradient approximation, N_i. Then the effective decision boundary feature matrix is estimated using the normal vectors as

$$\Sigma_{EDBFM} = \sum_i N_i N_i^T. \tag{2.7}$$

Next, the eigenvalue-eigenvector decomposition of the effective decision boundary feature matrix, Σ_{EDBFM}, is calculated and the normalised eigenvectors corresponding to non-zero eigenvalues are used as the transformation matrix from X to Y in (2.1).

3. Experiments

The data used in the experiment, the Anderson River data set, are a multi-source remote sensing and geographic data set made available by the Canada Centre for Remote Sensing (CCRS) [9]. The conjugate-gradient back-propagation (CGBP) algorithm with one hidden layer [10] was trained on the original data with 15, 20, 25, 30 and 35 hidden neurons. Each version of the CGBP network had 22 inputs and 6 outputs, and was trained six times with different initialisations. Then the overall average accuracies were computed for each version. With 30 hidden neurons 74.83% overall accuracy was reached for training data and 72.18% for test data. In comparison, the network with 35 hidden neurons gave an overall accuracy of 74.48% for training data and 72.18% for test data.

PCA was performed on the data. For the PCA approximately 99% of the variance in the data was preserved in 14 features, about 95% of the variance in 8 features, and 85% in 5 features. The CGBP with one hidden layer was trained on the PCA transformed data with a different number of input features. In each case, the number of hidden neurons was twice the number of input features, except for the 22 feature case where 30 hidden neurons were used. The classification results for the PCA are shown in Table 3.1. From Table 3.1 it can be seen that there was only about 1% decrease in overall training and test accuracies when 14 input features were used instead of 22. When less than 14 features were used, the classification accuracies decreased more significantly.

DA was then performed on the data. Since there were six information classes in the data, it was known that discriminant analysis would at maximum give only 5 input features using the criterion $\mathrm{tr}(\Sigma_W^{-1}\Sigma_B)$. According to this criterion approximately 97% of the variance was preserved in 4 features, and about 91% in 3 features. The CGBP with one hidden layer was trained on the DA transformed data with a different number of input features. In each case, the number of hidden neurons was twice the number of input features. The classification results for the DA are shown in Table 3.2. From Table 3.2 it can be seen that the use of DA for feature extraction resulted in

Table 3.1. Classification Accuracies for Principal Component Analysis

# of Features	Accumulated Eigenvalue	Overall Training Accuracy	Overall Test Accuracy
1	53.97	34.27	34.61
2	73.09	47.84	47.66
3	82.09	56.06	55.29
4	85.80	58.04	58.28
5	89.29	64.05	62.86
6	91.52	64.14	62.76
10	96.71	71.19	68.82
14	98.91	73.11	70.50
22	100.00	74.28	71.50

Table 3.2. Classification Accuracies for Discriminant Analysis

# of Features	Accumulated Eigenvalue	Overall Training Accuracy	Overall Test Accuracy
1	56.29	47.15	45.84
2	80.21	52.48	51.35
3	91.12	59.40	57.75
4	97.39	61.51	60.37
5	100.00	63.22	61.92

significantly less accurate classification by the neural network classifiers. The results are only comparable to the PCA results in Table 3.1 for five or fewer features. For those few features the DA results are slightly higher in terms of classification accuracies. However, the DA definitely suffered from only being able to give 5 total features.

Finally, DBFE was performed on the data. For the DBFE, approximately 99% of the variance in the data was preserved in 10 features, and about 94% in 6 features. The CGBP with one hidden layer was trained on the DBFE transformed data with a different number of input features. In each case, the number of hidden neurons was twice the number of input features, except for the 22 feature case where 30 hidden neurons were used. The classification results for the DBFE are shown in Table 3.3. From Table 3.3 it can be seen that there was only about 1% decrease in overall training and test accuracies when 10 input features were used instead of 22. Then, the performance decreased about 2% in terms of training and test accuracies when 6 features were used instead of 10. These results indicate that smaller neural network classifiers can be used to obtain similar results to the ones in the original feature space. Comparing the DBFE results in Table 3.3 to the corresponding PCA results in Table 3.1 it is clear that the DBFE always outperformed the PCA in terms of classification accuracies. For example, the accuracies for the DBFE with 6 features were similar to the PCA results with 10 features.

All the above results clearly demonstrate that the PCA is not an optimal input representation method for neural network classifiers. On the other hand, excellent classification results were achieved by using the DBFE.

Table 3.3. Classification Accuracies for Decision Boundary Feature Extraction

# of Features	Accumulated Eigenvalue	Overall Training Accuracy	Overall Test Accuracy
1	37.75	46.90	45.72
2	63.04	56.40	55.18
3	74.35	58.83	57.04
4	82.71	68.28	66.13
5	90.55	71.10	68.27
6	94.05	71.50	69.29
10	99.05	73.95	71.49
22	100.00	74.75	72.48

4. Conclusion

Neural networks were used to classify multi-source remote sensing and geographic data. Few feature extraction methods have been proposed for multi-source data but such data usually cannot be modelled by a simple multivariate statistical model. The DBFE method not only showed the best performance of the feature extraction methods in terms of classification accuracies when the dimensionality was reduced but also gave the best performance when the full 22 dimensional feature set was used. Since the DBFE algorithm does not assume any underlying probability density functions for the data, it takes full advantage of the distribution free nature of neural networks, and how neural network models define complex decision boundaries. With a reduced feature set, it is possible to obtain simpler classifiers but in the experiments these simpler classifiers gave similar accuracies to classifiers applied to the original data. The DBFE method also gives a feature matrix which has full rank, that is, as many features can be used as are in the original feature space. In contrast, discriminant analysis generally does not give a full rank matrix in the original feature space. Therefore, DA should be considered a less attractive method than DBFE for input representation for neural network classifiers.

With its good performance on the difficult data set, DBFE should both be considered an excellent feature extraction method and a desirable method for input data representation for neural networks.

References

1. K. Fukunaga, *Introduction to Statistical Pattern Recognition*, 2nd edition, Academic Press, NY, 1990.
2. J. Lampinen, and E. Oja, "Distortion Tolerant Pattern Recognition Based on Self-Organising Feature Extraction", *IEEE Transactions on Neural Networks*, vol. 6, pp. 539–547, 1995.

3. J. Mao, and A. K. Jain, "Artificial Neural Networks for Feature Extraction and Multivariate Data Projection", *IEEE Transactions on Neural Networks*, vol. 6, pp. 296–317, 1995.

4. E. Oja, "PCA, ICA, and Nonlinear Hebbian Learning", *Proceedings of the International Conference on Artificial Neural Networks (ICANN '95)*, held in Paris, France, on 9–13 Oct., pp. 89–94, 1995.

5. W. Fakhr, M. Kamel, and M. I. Elmasry, "The Adaptive Feature Extraction Nearest Neighbour Classifier", *Proceedings of the 1994 World Congress on Neural Networks*, vol. 3, pp. 123–128, 1994.

6. P. Blonda, V. la Forgia, G. Pasquaeriello, and G. Satalino, "Multispectral Classification by a Modular Neural Network Architecture", *Proceeding of the 1994 International Geoscience and Remote Sensing Symposium*, pp. 1873–1875, 1994.

7. B. Lee, D. Kim, Y. Cho, H. Lee, and H. Hwang, "Reduction of Input Nodes for Shift Invariant Second Order Neural Networks using Principal Component Analysis (PCA)", *Proceedings of the 1994 World Congress on Neural Networks*, vol. 3, pp. 144–149, 1994.

8. C. Lee, and D. A. Landgrebe, Decision Boundary Feature Extraction for "Non-Parametric Classifiers", *IEEE Transactions on Systems, Man and Cybernetics*, vol. 23, pp. 433–444, 1993.

9. D. G. Goodenough, M. Goldberg, G. Plunkett, and J. Zelek, "The CCRS SAR/MSS Anderson River Data Set", *IEEE Transactions on Geoscience and Remote Sensing*, vol. GE-25, pp. 360–367, 1987.

10. E. Barnard, "Optimisation for Training Neural Nets", *IEEE Transactions on Neural Networks*, vol. 3, pp. 232–240, 1992.

Spectral Pattern Recognition by a Two-Layer Perceptron:
Effects of Training Set Size

Petra Staufer and Manfred M. Fischer*

Department of Economic and Social Geography,
Vienna University of Economics & BA, A-1090 Vienna, Augasse 2-6.
e-mail: petra@wigeo1.wu-wien.ac.at, manfred.m.fischer@wu-wien.ac.at

Summary. Pattern recognition in urban areas is one of the most challenging issues in classifying satellite remote sensing data. Parametric pixel-by-pixel classification algorithms tend to perform poorly in this context. This is because urban areas comprise a complex spatial assemblage of disparate land cover types – including built structures, numerous vegetation types, bare soil and water bodies. Thus, there is a need for more powerful spectral pattern recognition techniques, utilising pixel-by-pixel spectral information as the basis for automated urban land cover detection. This paper adopts the multi-layer perceptron classifier suggested and implemented in [5]. The objective of this study is to analyse the performance and stability of this classifier - trained and tested for supervised classification (8 a priori given land use classes) of a Landsat-5 TM image (270 × 360 pixels) from the city of Vienna and its northern surroundings - along with varying the training data set in the single-training-site case. The performance is measured in terms of total classification, map user's and map producer's accuracies. In addition, the stability with initial parameter conditions, classification error matrices, and error curves are analysed in some detail.

1. Introduction

The remote sensing literature on neural network applications to multispectral pattern recognition is relatively new, dating back about six to seven years. The first studies established the feasibility of back-propagation neural classifiers (see [1, 9, 4]). Subsequent studies analysed this type of classifier in more detail and compared it to current best practice such as the Gaussian maximum likelihood (see, e.g., [2, 12], [5]). One of the most difficult aspects of supervised classification is the choice of representative training sites for each land cover class (see, e.g., [9, 6]). Ideally a good classifier is one which after training with the feature values from the example classes is capable of being used to map or collect statistics accurately over much wider areas of territory from the remotely sensed data without the need for further ground survey.

Time and expenses can be saved if smaller samples of training pixels are required for obtaining a sufficient generalisation performance and stability

* The authors gratefully acknowledge a grant from the Austrian Ministry of Science and Arts, GZ 308.937/2-IV/3/95.

with initial parameter conditions of the classifier. The motivation for the current study is to investigate training and generalisation behaviour dependent on training set sizes for the single training site case. In-sample and out-of-sample performances are measured in terms of total classification, map user's and map producer's accuracies. In addition, classification error matrices are used to track the results by a close and systematic examination.

In the following section, we briefly describe, first, the pattern recognition task at hand as a supervised multispectral pixel-by-pixel classification problem, second, the major characteristics of the two-layer perceptron classifier used, third, the performance measures and the data sets for training and testing. In section 3, the experimental set up and the estimation procedure are outlined, before discussing the results obtained. Particular emphasis is laid here in this study on the crucial issue of the generalisation behaviour of the MLP-1 classifier as suggested by [5] depending on various training data sets. Some conclusions are drawn in the final section.

2. The Pattern Recognition Problem and the Data

The objective of satellite image classification operations is to replace visual analysis of the image data with quantitative techniques for automating the identification of man-made objects as well as natural phenomena in a scene. This normally involves the analysis of multispectral data and the application of statistical decision rules for determining the land cover identity for each pixel of an image, based on the spectral radiances observed in the data. Computational neural networks (CNNs, or simply, NNs) represent a different approach to the pattern recognition problem, as they do not rely explicitly on the probabilistic nature both of the information to be processed, and of the form in which to express the results. The role of neural networks is to provide general semi-parametrised non-linear mappings between multidimensional spaces (see, e.g., [3]), i.e., NNs can be viewed as adaptive model-free function estimators using a non algorithmic strategy.

The problem considered here is a typical *supervised multispectral pixel-by-pixel classification* problem using urban landcover information in which the classifier is trained with pre-selected examples of the landuse classes to be recognised in the data set. The problem to discriminate between eight urban landuse categories, as outlined in table 2.1, is challenging because urban agglomerations embrace a complex composition of spectral landcover types, including built structures like (sub)urban settlement areas, parks and recreation areas with water bodies and numerous types of vegetation. These categories are meaningful to photointerpreters and landuse mappers, but are not necessarily spectrally homogeneous — a problem hard to tackle by conventional spectral pixel-by-pixel classification techniques (see, e.g., [7]).

This problem has been tackled by means of the two-layer perceptron MLP-1 with 14 logistic hidden units and softmax output activation suggested and implemented in [5]. Six input units represent the spectral bands of the multispectral image and eight output units the a priori given landuse classes (see table 2.1). The weight elimination pruning strategy has been used to control for model complexity, i.e., the size of the hidden layer. The network was trained utilising the most simple local optimisation technique based on gradient descent in order to minimise the least mean squared error (LASE) performance measure. The parameter estimation is stochastic epoch based (epoch size 3), the update for the weight parameter $\omega_{rs}^{(n)}$ that connects the n-th node of the $(n-1)$-th layer to the j-th node of the $(n+1)$-th layer at step (t) is given by

$$\omega_{rs}^{(n)}(t+1) = \omega_{rs}^{(n)}(t) + \eta \frac{\partial E}{\partial \omega_{rs}^{(n)}} \qquad n = 1, 2 \qquad (2.1)$$

where E denotes the standard LASE function to be minimised over the set of training examples, and η the learning rate set to 0.8. For a detailed description of the network set up procedure see [5], the statistical approach to learning in such networks is described in [8]. Since all iterative procedures are sensitive to different starting points, it is important to perform several random runs. We deliberately have chosen five different runs with initial weights drawn at random from a uniform distribution in $[-0.1, +0.1]$.

A common means for expressing *classification performance* is the preparation of confusion matrices. This involves both in-sample (training) and especially out-of-sample classification performance. The following standard measures such as the *classification error matrix* or *confusion matrix* f_{lk} with $f_{lk}(l, k = 1, \ldots, C)$ listing the pixels assigned by the classifier to category k versus the known landuse (ground truth) category l, the *map user's accuracy*, ν_k, for the ground truth category $k = 1, \ldots, C$, the *map producer's accuracy* π_l for the classifier's category $l = 1, \ldots, C$, and the *total classification accuracy* τ (or the total classification error $\tau\prime$ defined as $\tau\prime = (100 - \tau)$). A less common measure is the *KHAT statistics*, $\hat{k}$, measuring the actual agreement between ground truth category l and classifier category k and the chance agreement between ground truth category and a random classifier category. It incorporates the non diagonal elements of the error matrix as a product of row and column marginal. One of the principal advantages of $\hat{k}$ statistics is the ability to use this value as a basis for determining the difference among matrices (see, e.g., [10]) and hence serves as an appropriate measure in the current experimental setup.

For the purpose of experimentation, a Landsat-5 TM image covering the city of Vienna and its northern surroundings was selected (270 x 360 pixels; TM Quarter Scene 190-026/4; location of the centre: $16°23'E$, $48°14'N$; observation date: June 5, 1985). Familiarity with the area allowed for accurate class training and test site selection. Additional reference data was provided through a series of analogue orthophotos gathered during the same year, and a parcel-based landuse map of the city of Vienna. The six spectral bands of the Landsat sensor with a ground resolution of $30m$ x $30m$ were used for classification whereas the thermal band (TM channel 6) was not considered for this task. A single training site has been selected for each of the eight categories chosen to cover the majority of urban landuse features in the Vienna image (see table 2.1). This resulted in a database consisting of 2,460 pixels (about 2.5 percent of all the pixels in the image) that are described by six-dimensional feature vectors and their class membership (target values).

Table 2.1. Categories Used for Landuse Classification and Number of Pixels

Category Number	Description of the Category	Pixels per Site
C1	Mixed grass and arable farmland	250
C2	Vineyards and areas with low vegetation cover	427
C3	Asphalt and concrete surfaces (industrial/commercial/traffic areas)	192
C4	Woodland and public gardens with trees	602
C5	Low density residential and industrial areas (suburban)	154
C6	Densely built up residential areas (urban)	444
C7	Water courses	230
C8	Stagnant water bodies	161
	Total Number of Pixels for Training and Testing	2,460

In figure 2.1, scattergrams for the imagebands 1-5 and 7 show the actual pixel distribution of the full training sample of 2,460 feature vectors. First, the image bands TM1, TM2 and TM3 show a high correlation, the clusters for all landuse classes except class 3 and 4 are overlapping or contain each other. Band 4, 5 and 6 indicate a better separability. Second, there is some confusion between densely built up areas and water bodies, which is peculiar. The water body in this case is the River Danube that flows through the city and is surrounded by densely built up areas. The confusion could be caused by a "boundary problem" where there are mixed pixels at the boundary.

A few of the categories, such as 'suburban', are sparsely distributed in the image. Thus, in order to keep the classification accuracy calculations from being dominated by a few of the more prevalent classes, the database has been divided into a training set (two thirds of the training site pixels) and a testing set (one third of the training site pixels) by stratified random sampling, stratified in terms of the a priori probability of class occurrence of the eight categories (see, e.g., [11]). This resulted into 1,640 training and 820 testing pixels. Table 2.2 shows the mean grey scale values over the six TM-bands and the average standard deviation ($\bar{\sigma}$) of each landuse category as defined by this training sample. Notice, higher $\bar{\sigma}$-values characterise complex landuse classes and hence, indicate a larger overlap of spectral class boundaries.

Four further training sets were produced by reduction of the basic training samples of the eight categories whereas the testing set remained the same. Training set 2 represents 80% (a total of 1,313 pixels) of the basic training set, training set 3 includes 60% (984 pixels), training set 4 40% (656 pixels), and training set 5 only 20% (327 pixels). Since the resampling procedure was randomly stratified according to the a priori probability of class occurrence

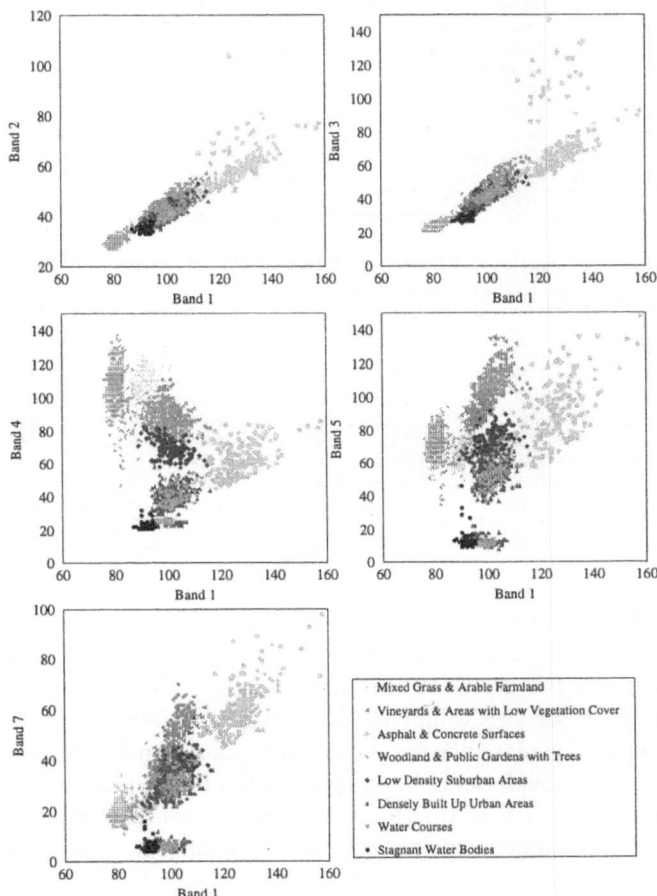

Fig. 2.1. Scattergrams of the Grey Value Distribution per Landuse Class

as well, the number of training pixels for sparsely distributed landuse classes became rather small. The basic statistical characteristics of the additional training sets (see table 2.3) are comparable to those of the basic training set 1. This should guarantee a fair comparison of the classifier's performance and evaluation of the impact of training set size.

3. The Classifier and the Simulation Experiments

The main objective of our experimental setup is to analyse the *stability of in-sample and out-of-sample classification performance* of the MLP-1 classi-

Table 2.2. Statistical Characteristics of the Basic Training and Testing Set

Category	Training Set 1			Testing Set		
Number	Pixels	Mean	$\bar{\sigma}$	Pixels	Mean	$\bar{\sigma}$
C1	167	59.92	6.14	83	61.66	6.98
C2	285	70.56	7.83	142	71.94	7.99
C3	128	77.88	10.23	64	82.03	15.72
C4	402	56.57	5.72	200	54.69	7.73
C5	102	59.60	5.28	52	62.13	4.31
C6	296	49.51	8.95	148	46.70	9.98
C7	153	43.09	10.76	77	42.45	10.71
C8	107	32.62	1.77	54	32.67	2.42
Total	1,640	56.98	7.61	820	57.00	9.07

Table 2.3. Basic Statistical Characteristics of the Training Sets 2 – 5

Cat.	Training Set 2			Training Set 3			Training Set 4			Training Set 5		
	Pixels	Mean	$\bar{\sigma}$	Pixels	Mean	$\bar{\sigma}$	Pixels	Mean	$\bar{\sigma}$	Pixels	Mean	$\bar{\sigma}$
C1	134	60.00	6.12	100	59.76	5.88	67	60.17	6.56	33	59.59	6.24
C2	285	70.59	7.85	171	70.58	8.01	114	70.52	6.61	57	70.43	7.96
C3	102	77.84	10.57	77	77.89	10.19	51	77.85	10.35	26	78.02	8.87
C4	322	56.12	5.81	241	55.92	5.68	161	56.29	5.79	80	55.84	5.37
C5	82	59.34	5.34	61	59.92	5.46	41	59.14	5.02	20	60.68	4.91
C6	237	49.82	8.38	178	50.00	8.94	118	48.75	8.91	59	48.27	10.83
C7	122	42.96	10.83	92	42.86	10.41	61	43.44	11.35	32	43.62	10.59
C8	86	32.57	1.74	64	32.36	1.02	43	33.00	2.45	21	32.81	1.95
Total	1,313	57.02	7.61	984	57.00	7.53	656	56.94	7.64	327	56.82	7.64

fier with respect to *varying training set sizes*. The experimental setup was led by the hypothesis that the smaller training sets should result in poorer performance for the classifiers. In other words, it is expected that a reduction in training set size would yield a significant decrease of generalisation accuracy, since the classifier requires an adequate number of samples in each category to describe decision boundaries in the feature space. Intuitively one might expect furthermore that a neural classifier would require more samples for spectrally heterogeneous landuse categories since it assumes no statistical distribution and thus would need more information to define these decision regions.

Another important aspect for real world applications of neural classifiers is their *stability with initial parameter conditions*. The objective function of gradient descent based multilayer perceptron networks has multiple local minima and therefore this network type is known to be sensitive to details of initial weight values. In our experiments the network topology of the MLP-1 classifier is fixed (6:14:8), and so is the number of 196 free parameters (adjustable weights) and the gradient descent control term $\eta = 0.8$. We used five different sets of initial weights which were chosen from a uniform random distribution in $[-0.1, +0.1]$ to investigate the stability aspects of the classifier in the setting at hand.

All the simulations described are performed using the epoch based stochastic version of back-propagation, where the weights are updated after each epoch of three randomly chosen patterns in the training sets (epoch size 3). The spectral grey scale values of the six TM-bands were transformed in $[0, 1]$

Table 3.1. Effects of Varying Training Sets and Different Initial Parameter Conditions

	Total In-Sample Accuracy in %					Total Out-Of-Sample Accuracy in %				
	Set 1	Set 2	Set 3	Set 4	Set 5	Set 1	Set 2	Set 3	Set 4	Set 5
τ_1 (Run 1)	91.89	91.32	91.77	90.24	85.02	**90.00**	87.20	88.29	86.95	83.29
τ_2 (Run 2)	91.40	91.32	91.77	90.24	84.71	89.02	87.20	88.29	87.32	84.63
τ_3 (Run 3)	91.89	91.01	91.77	89.18	79.82	89.51	**88.66**	87.56	**88.17**	76.46
τ_4 (Run 4)	91.46	91.55	91.97	90.40	83.49	88.78	88.41	**88.54**	87.32	82.68
τ_5 (Run 5)	91.52	90.86	90.24	90.40	86.24	88.54	86.59	86.95	87.32	**85.73**
$\varrho^\tau_{(1-5)}$	0.49	0.69	1.73	1.22	6.42	1.46	2.07	1.59	1.22	9.27
$\bar{\tau}_{(1-5)}$	91.63	91.21	91.50	90.09	83.86	89.17	87.61	87.93	87.42	82.56
$\sigma^\tau_{(1-5)}$	0.23	0.27	0.72	0.52	2.46	0.59	0.88	0.66	0.45	3.61

	In-Sample $\hat{k}$-Value					Out-Of-Sample $\hat{k}$-Value				
	Set 1	Set 2	Set 3	Set 4	Set 5	Set 1	Set 2	Set 3	Set 4	Set 5
$\hat{k}_1$ (Run 1)	0.90	0.90	0.90	0.89	0.83	0.88	0.85	0.86	0.85	0.81
$\hat{k}_2$ (Run 2)	0.90	0.90	0.91	0.89	0.82	0.87	0.85	0.86	0.85	0.82
$\hat{k}_3$ (Run 3)	0.90	0.89	0.90	0.87	0.77	0.88	0.87	0.86	0.86	0.74
$\hat{k}_4$ (Run 4)	0.90	0.90	0.91	0.88	0.81	0.87	0.87	0.87	0.85	0.80
$\hat{k}_5$ (Run 5)	0.90	0.89	0.89	0.89	0.84	0.87	0.85	0.85	0.85	0.83
$\varrho^{\hat{k}}_{(1-5)}$	0.00	0.01	0.02	0.02	0.07	0.01	0.02	0.02	0.01	0.09
$\widehat{\bar{k}}_{(1-5)}$	0.90	0.90	0.90	0.88	0.81	0.87	0.86	0.86	0.85	0.80
$\sigma^{\hat{k}}_{(1-5)}$	0.00	0.005	0.008	0.009	0.027	0.005	0.011	0.007	0.004	0.035

for mapping the observed signals onto a set of input unit activations. The softmax activation function for the final layer of MLP-1 generates output values in the range $[0, 1]$ which can be interpreted as probabilities of class membership, conditioned on the outputs of the hidden units. The output unit with the maximum activation indicates the actual mapping of an input pixel onto a specific output class C. All classifiers were trained for 10 cycles where one cycle is defined as a full presentation of the entire training set. For the purpose of monitoring the estimation and generalisation behaviour of the classifiers the performance measures described in section 2 were computed after each training cycle.

Table 3.1 gives a summary of the total in-sample and out-of-sample classification performance of the MLP-1 classifier trained with the five different training sets in five runs with different initial sets of weights. The upper part of the table shows the total classification accuracy $\tau_1, \ldots, \tau_5$ in percent. The bold letters indicate the classifiers with the best generalisation capabilities. At the bottom, the corresponding $\hat{k}$-values are listed. The major results confirm the hypothesis that *in-sample and out-of-sample performances* measured in terms of total classification accuracy of the MLP-1 classifier *generally increase with increasing training set size*. The stability of the classifier's performance measured in terms of the range $\varrho^\tau_{(1-5)}$ and standard deviation $\sigma^\tau_{(1-5)}$ of the total classification accuracy over the the five different random parameter initialisations (run 1-5) *generally increases with training set size* as well. The range $\varrho^\tau_{(1-5)}$ (and the standard deviation $\sigma^\tau_{(1-5)}$) of the generalisation accuracy decrease from 9.27 (3.61) for MLP-1 trained with 327 pixels (training set 5) to 1.46 (0.59) for MLP-1 trained with 1,640 pixels (training set 1). The k-values, which incorporate the nondiagonal elements of the error matrices as a product of row and column marginal, serve as an indicator of the extent to which the percentage correct values of an error matrix are due to "true" agreement versus "chance" agreement. The calculated average $\hat{k}$-values, $\widehat{\overline{k}}_{(1-5)}$, vary from 0.87 for MLP-1 trained with 1,640 pixels (training set 1) to 0.80 for MLP-1 trained with 327 pixels (training set 5) indicate that the observed classifications are 87% (80%, respectively) better than those resulting from a random assignment.

Figure 3.1 shows a mapping of the "cost of performance and stability" in terms of the number of training pixels versus the "gain" in terms of the average generalisation accuracy, $\overline{\tau}_{(1-5)}$, and the stability indicated by the standard deviation over the five runs, $\sigma^{\tau}_{(1-5)}$. $\overline{\tau}_{(1-5)}$ decreases from 89.17% achieved by training the MLP-1 classifier with 1,640 patterns (training set 1) to 82.56% using just 327 patterns for training. The cost for 6.61% average out-of-sample performance gain is a five times larger training set, and hence, the collection of five times more ground truth information. The knowledge obtained through that cost-performance-analysis should have practical consequences in a way as we could show that a deliberate design of a simple MLP network classifier together with a small number of training patterns yields a sufficient generalisation performance.

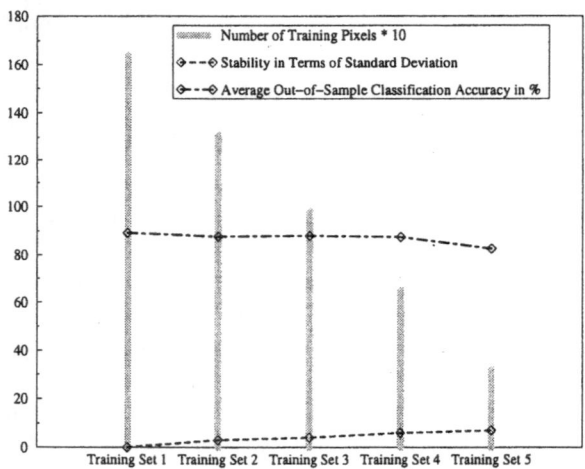

Fig. 3.1. The Cost of Performance and Stability: Does That Pay?

A closer examination of the error matrices for the best performing runs per training set (see tables 3.2, 3.3, 3.4, 3.5, 3.6) and the calculation of map user's and map producer's classification accuracies (ν_k and π_l) shows that complex landuse classes, such as low density residential and industrial suburban areas (class 5) and densely built up urban areas (class 6) may have significantly lower ν_k- and π_l-values than spectrally homogeneous landuse

Table 3.2. Out-of-Sample Confusion Matrix for MLP-1 Using Training Set 1

Classifier's Category	Ground Truth Category								
	Cl. 1	Cl. 2	Cl. 3	Cl. 4	Cl. 5	Cl. 6	Cl. 7	Cl. 8	ν_k
Class 1	79	4	0	0	0	0	0	0	95.18
Class 2	0	142	0	0	0	0	0	0	100.00
Class 3	0	0	64	0	0	0	0	0	100.00
Class 4	2	3	0	193	1	0	0	1	96.50
Class 5	0	6	0	0	46	0	0	0	88.46
Class 6	0	0	0	0	0	113	31	4	76.35
Class 7	0	0	0	0	0	29	48	0	62.34
Class 8	0	0	0	0	0	1	0	53	98.15
π_l	97.53	91.61	100	100	97.87	79.02	60.76	91.38	

Table 3.3. Out-of-Sample Confusion Matrix for MLP-1 Using Training Set 2

Classifier's Category	Ground Truth Category								
	Cl. 1	Cl. 2	Cl. 3	Cl. 4	Cl. 5	Cl. 6	Cl. 7	Cl. 8	ν_k
Class 1	77	5	0	1	0	0	0	0	84.34
Class 2	0	136	0	0	6	0	0	0	98.59
Class 3	0	0	63	0	1	0	0	0	100.00
Class 4	1	1	0	194	3	0	0	1	94.50
Class 5	0	0	1	0	51	0	0	0	94.23
Class 6	0	0	0	0	1	104	33	10	77.70
Class 7	0	0	0	0	0	29	48	0	62.34
Class 8	0	0	0	0	0	0	0	54	98.15
π_l	97.72	95.77	98.44	99.49	82.26	78.20	59.26	83.08	

classes. For example, the map user's accuracy for class 5 shows a maximum of 94.23 percent for MLP-1 using training set 2 and a minimum of 65.38 percent (set 5). The corresponding map producer's accuracies are 82.26 percent and 97.14 percent, respectively. The $\nu_k(\pi_l)$-values for class 6 lie between 69.59 (76.87) percent (set 3) and 79.73 (77.12) percent (set 4). Second, there is a large proportion of cross-assigned pixels between class 6 and class 7 (water courses). In that case the $\nu_k(\pi_l)$-values have their minimum at 55.55 (61.76) percent (set 4), the other training sets come up with 62.34 percent ν_k- and slightly different π_l-values (57.83–61.54 percent). The confusions are likely to be mixed pixels effects caused by spectral signatures of pixels covering regions of diverse landcover categories. In both cases, the problem of mixed pixels is a severe one – it occurs if the average size of the regions of homogeneous spectral signature is not much larger than the pixel size. The integration of spatial information (e.g., texture), sub-pixel information (e.g., spectral mixture analysis), advanced sensor technology (e.g., higher geometric resolution with a constant or even improved number spectral bands), and/or ancillary information from GIS databases might be appropriate to solve these problems which cannot be tackled by that simple MLP-1 classifier working in a pure spectral pixel-by-pixel information context.

The *analysis of the generalisation ability* of the best performing trials of MLP-1 using different training sets completes the investigations. Figure 3.2 shows the out-of-sample classification error curves as a function of training time. It is clear, that different training set sizes can lead to more or less major differences in the starting stage of the training and generalisation process.

Table 3.4. Out-of-Sample Confusion Matrix for MLP-1 Using Training Set 3

Classifier's Category	Cl. 1	Cl. 2	Cl. 3	Cl. 4	Cl. 5	Cl. 6	Cl. 7	Cl. 8	ν_k
Class 1	78	5	0	0	0	0	0	0	93.98
Class 2	0	142	0	0	0	0	0	0	100.00
Class 3	0	2	61	0	0	1	0	0	95.31
Class 4	1	2	0	194	2	0	0	1	97.00
Class 5	0	5	0	0	47	0	0	0	90.38
Class 6	0	0	0	0	1	103	35	9	69.59
Class 7	0	0	0	0	0	29	48	0	62.34
Class 8	0	0	0	0	0	1	0	59	98.15
π_l	98.73	91.03	100.00	100.00	94.00	76.87	57.83	84.13	

Table 3.5. Out-of-Sample Confusion Matrix for MLP-1 Using Training Set 4

Classifier's Category	Cl. 1	Cl. 2	Cl. 3	Cl. 4	Cl. 5	Cl. 6	Cl. 7	Cl. 8	ν_k
Class 1	79	4	0	0	0	0	0	0	95.18
Class 2	8	134	0	0	0	0	0	0	94.37
Class 3	0	0	64	0	0	0	0	0	100.00
Class 4	6	2	0	191	0	0	0	1	95.50
Class 5	0	7	3	0	42	0	0	0	80.77
Class 6	0	0	0	0	0	118	26	4	79.73
Class 7	0	0	0	0	0	35	42	0	54.55
Class 8	0	0	0	0	0	1	0	53	98.15
π_l	84.95	91.16	95.52	100.00	100.00	77.12	61.76	91.38	

Table 3.6. Out-of-Sample Confusion Matrix for MLP-1 Using Training Set 5

Classifier's Category	Cl. 1	Cl. 2	Cl. 3	Cl. 4	Cl. 5	Cl. 6	Cl. 7	Cl. 8	ν_k
Class 1	76	6	0	1	0	0	0	0	91.57
Class 2	5	137	0	0	0	0	0	0	96.48
Class 3	0	1	63	0	0	0	0	0	98.44
Class 4	10	2	0	185	0	0	0	3	92.50
Class 5	0	17	1	0	34	0	0	0	65.38
Class 6	0	0	0	0	1	106	30	11	71.62
Class 7	0	0	0	0	0	29	48	0	62.34
Class 8	0	0	0	0	0	0	0	54	100.00
π_l	83.52	84.05	98.44	99.46	97.14	78.52	61.54	79.41	

After six training cycles the differences between MLP-1 trained with training set 1, 2 and 3 more or less vanish and the classifiers tend to converge after the fully stochastic presentation of 9,840 (set 1), 7,878 (set 2), and 5,904 training pixels (set 3), respectively. In contrast, the smaller numbers of training pixels in set 4 (656 pixels) and set 5 (327 pixels) lead to oscillation of generalisation performance under constant parameter conditions. As suggested in [5], a variable learning rate adjustment (declining learning rate) might lead to a more stable generalisation behaviour and slightly better performance.

4. Conclusions and Outlook

This study has focussed on the crucial issue of the generalisation (out-of-sample classification performance) of the MLP-1 classifier implemented in [5]

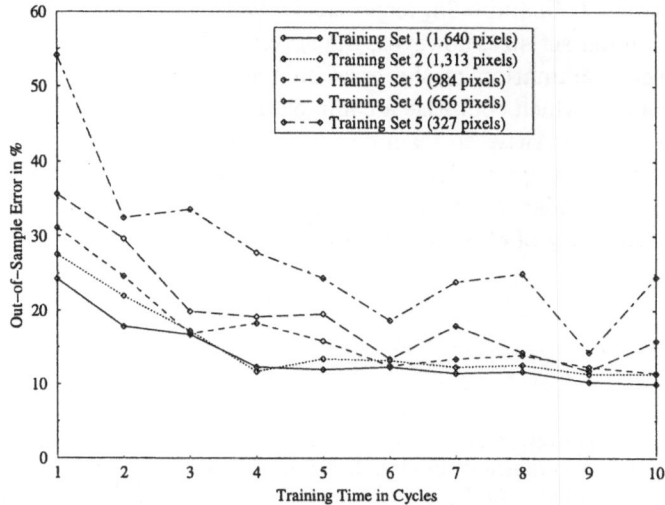

Fig. 3.2. Out-of-Sample Error Decrease as a Function of Training Time

with particular emphasis on the important question whether there are some regularities in dependence of the training data set size in the single training site case. For real world applications of neural classifiers in remote sensing both factors are equally relevant. First, the size of the training set, and hence, the number of ground truth pixels to collect, is highly important under tense time and financial restrictions. Second, the stability of out-of-sample performance and the reliability of neural classifier systems under given conditions has to be proven to illustrate their superiority over conventional parametric techniques.

The experimental results clearly indicate that the generalisation performance measured in terms of total classification accuracy generally increases with increasing training set size. The elasticity of generalisation performance – in other words, the relative impact of training set size on performance gain – was found to be surprisingly low. For e.g., in our experimental set up 40% less training pixels yielded a out-of-sample performance loss of just 1.24% from an average performance of 89.17% to 87.93%, and almost 80% less training pixels resulted in a loss of 6.61% to 82.56%, respectively.

The second important finding is that the stability of the classifier's performance measured in terms of the range $\varrho^\tau_{(1-5)}$ and standard deviation $\sigma^\tau_{(1-5)}$ of total classification accuracy over five different parameter initialisations $(\tau_1, \ldots, \tau_5)$ generally increases with increasing training set size as well. It is worthwhile to mention that a reduction of the number of training pixels down to 40% of the initial size had almost no influence on the stability of results. The figures for the range of the generalisation accuracy measured in terms of $\varrho^\tau_{(1-5)}$ vary between 1.22 and 2.07, and for the standard deviation

$\sigma^T_{(1-5)}$ between 0.45 and 0.88, respectively. With the reduction to 20% of the original training set size, however, the stability figures changed significantly.

A closer examination of the error matrices shows that especially those landuse classes, which comprise a complex assemblage of disparate landcover types, are more sensitive to training set size. This leads to the recommendation that the stratification rule for random sampling of very small training sets should be guided by the complexity of landuse classes as well as by the a priori probability of class occurrence.

References

1. J. A. Benediktsson, P. A. Swain, and O. K. Ersoy, "Neural network approaches versus statistical methods in classification of multisource remote sensing data", *IEEE Transactions on Geoscience and Remote Sensing*, vol. 28, pp. 540–551, 1990.
2. H. Bischof, W. Schneider, and A. J. Pinz, "Multispectral classification of Landsat-images using neural networks", *IEEE Transactions on Geoscience and Remote Sensing*, vol. 30, no. 3, pp. 482–490, 1992.
3. C. M. Bishop, *Neural Networks for Pattern Recognition*, Oxford: Clarendon Press, 1995.
4. D. L. Civco, "Artificial neural networks for land-cover classification and mapping", *International Journal of Geographical Information Systems*, vol. 7, no. 2, pp. 173–186, 1993.
5. M. M. Fischer, S. Gopal, P. Staufer and K. Steinnocher, "Evaluation of Neural Pattern Classifiers for a Remote Sensing Application", *Geographical Systems*, vol. 4, no. 1, 1997 [in press].
6. G. M. Foody, M. B. McCulloch, and W. G. Yates, "Classification of Remotely Sensed Data by an Artificial Neural Network: Issues Related to Training Data Characteristics", *Photogrammetric Engineering and Remote Sensing*, vol. 61, no. 4, pp. 391–401, 1995.
7. T. Fung and K. Chang, "Spatial composition of spectral classes: a structural approach for image analysis of heterogeneous land-use and land-cover types", *Photogrammetric Engineering and Remote Sensing*, vol. 60, no. 2, pp. 173–180, 1994.
8. S. Gopal and M. M. Fischer, "Learning in single hidden layer feedforward neural network models", *Geographical Analysis*, vol. 28, no. 1, pp. 38–55, 1996.
9. G. F. Hepner, T. Logan, N. Ritter and N. Bryant, "Artificial neural network classification using a minimal training set: Comparison to conventional supervised classification", *Photogrammetric Engineering and Remote Sensing*, vol. 56, no. 4, pp. 469–473, 1990.
10. T. M. Lillesand, and R. W. Kiefer, *Remote Sensing and Image Interpretation*, New York et al: John Wiley & Sons, 1994.
11. A. H. Strahler, "The use of prior probabilities in maximum likelihood classification of remotely sensed data", *Remote Sensing of Environment*, vol. 10, pp. 135–163, 1980.
12. G. G. Wilkinson, F. Fierens, F. and I. Kanellopoulos, "Integration of neural and statistical approaches in spatial data classification", *Geographical Systems*, vol. 2, no. 1, pp. 1–20, 1995.

Comparison and Combination of Statistical and Neural Network Algorithms for Remote-Sensing Image Classification

Fabio Roli, Giorgio Giacinto, and Gianni Vernazza

Department of Electrical and Electronic Engineering,
University of Cagliari,
Piazza D'Armi, 09123, Cagliari, Italy.
e-mail {roli,giacinto,vernazza}@diee.unica.it

Summary. In recent years, the remote-sensing community has became very interested in applying neural networks to image classification and in comparing neural networks performances with the ones of classical statistical methods. These experimental comparisons pointed out that no single classification algorithm can be regarded as a "panacea". The superiority of one algorithm over the other strongly depends on the selected data set and on the efforts devoted to the "designing phases" of algorithms. In this paper, we propose the use of "ensembles" of neural and statistical classification algorithms as an alternative approach based on the exploitation of the complementary characteristics of different classifiers. Classification results provided by image classifiers contained in these ensembles are "merged" according to statistical combination methods. Experimental results on a multi-sensor remote-sensing data set point out that the use of classifiers ensembles can constitute a valid alternative to the development of new classification algorithms "more complex" than the present ones. In particular, we show that the combination of results provided by statistical and neural algorithms provides classification accuracies better than the ones obtained by single classifiers after long "designing phases".

1. Introduction

Classification of remote-sensing images has traditionally been performed by classical statistical methods (e.g., bayesian and k-nearest-neighbour classifiers). In recent years, the remote-sensing community has become very interested in applying neural networks to image classification and in comparing neural networks performances with the ones of statistical methods [1, 2]. Experimental comparisons among neural and statistical classifiers reported in the remote-sensing literature pointed out that no single classification algorithm can be regarded as a "panacea". In particular, it seems to us that the reported superiority of one algorithm over the other strongly depends on the selected data set and on the effort devoted to the "designing phase" of each classifier (i.e., designing of classifier "architecture", choice of learning parameters, etc.). As an example, the superiority of the k-nearest-neighbour classifier (KNN) over the Multilayer Perceptron (MLP) neural network, or vice versa, strongly depends on the efforts devoted to the related designing phases: selection of the appropriate "k" value and of the appropriate "distance measure" for the KNN classifier, selection of the appropriate architecture and suitable learning parameters for the MLP. In addition, according to our experience, any algorithm may reach a certain level of classification accuracy through a reasonable "designing" effort. Further improvements often require an increasingly expensive designing phase [2, 3, 4].

In spite of the above considerations, most of the present research work on remote-sensing image classification is focused on the development of new statistical and neural classification algorithms. No emphasis is given to the exploitation of the complementary characteristics of existing algorithms by suitable techniques that "combine" results provided by different classifiers. Few papers addressed the need of integrating classification results provided by different classification algorithms [5]. In addition, no investigation has been carried out to evaluate the benefits of combining different classification algorithms in order to reduce the complexity of the designing phase (i.e., in order to obtain the desired classification accuracy with a reduced designing effort).

In this paper, we propose the use of "classifiers ensembles" as an alternative approach based on the exploitation of the complementary characteristics of different classification algorithms. Ensembles of neural and statistical classifiers are considered. We report experimental results on a multi-sensor remote-sensing data set that prove that the use of classifiers ensembles can constitute a valid alternative to the development of new algorithms "more complex" than the present ones. In particular, we show that the combination of results provided by statistical and neural algorithms provides classification accuracies better than the ones obtained by single classifiers after long "designing" phases.

2. Comparison of Statistical and Neural Network Classifiers

The first attempts to apply MLP neural networks to the classification of remote-sensing data were described by Decatur [6]. He compared MLP performances with the ones of a Bayesian classifier for the classification of terrain radar images. He found that significant improvements can be obtained by the MLP classifier. Benediktsson at al. applied MLPs to classification of multi-source remote-sensing data (in particular, they used LANDSAT MSS and topographic data) [1]. Classification performances were compared with the ones of a statistical parametric method which takes into account the relative reliabilities of the sources of data. They concluded that the relative performances of the two methods mainly depend on prior knowledge about the statistical distribution of data. MLPs are appropriate for cases where such distributions are unknown, for they are "data-distribution free". The use of an MLP for cloud classification using LANDSAT MSS images was described by Lee et al. [7]. They proved that very high accuracy of cloud classification can be obtained by using a four-layer perceptron. Bischof et al. reported the application of a three-layer perceptron for classification of LANDSAT TM data [8]. They showed that the MLP performs better than a Bayesian classifier, and that textural information can be integrated into the neural network classifier without the explicit definition of a texture measure. Azimi-Sadjadi et al. developed a structured neural network to classify radar images [9]. The network architecture consists of four subnets (each for a different polarisation of a radar signal) and of a final network which combines the subnet outputs to accomplish the decision task. The reported results proved the advantages of a "combined polarisation architecture" that exploits the peculiarities of radar data. Salu and Tilton introduced a new neural network model, named the Binary Diamond Neural Network, for the classification of LANDSAT 4 TM data [10]. The Binary Diamond is a multilayer, feed-forward neural network that learns from examples, in the "one shot" mode. The reported results show that the Binary Diamond neural network performs much better than MLPs. The only drawback of this neural model seems to be the considerable use of memory.

The experimental results reported in the above-mentioned papers pointed out that neural networks usually perform better than parametric statistical classifiers like the Gaussian classifier. On the other hand, neural networks provide classification accuracies similar to the ones obtained by non-parametric classifiers like the k-nearest-neighbour classifier. These achievements are in agreement with the theory of statistical pattern recognition [11], as neural networks like MLPs are a particular kind of non parametric classifier. In addition, according to our experience [2, 3, 4], any algorithm may reach a certain level of classification accuracy through a reasonable "designing" effort. Further improvements often require an increasingly expensive designing phase. Consequently, it seems to us that improvements in classification accuracy might be gained more easily by exploiting complementary characteristics of different classification algorithms. This objective can be reached by "combining" classification results provided by multiple classifiers.

3. Methods for Combining Multiple Classifiers

In the following, we propose various methods that can be used to combine results provided by an ensemble of classification algorithms (e.g., an ensemble of neural networks). In particular, some combination methods previously proposed in the handwriting recognition field are described [12]. These methods assume that classifiers contained in the ensemble behave "independently", that is, they make uncorrelated classification mistakes.

Let us assume an image classification problem with M "data classes". Each class represents a set of specific patterns. Each pattern is characterised by a feature vector $\underline{X}$. In addition, let us assume that K different classification algorithms are available to solve the classification problem at hand. Therefore, we can consider ensembles formed by "k" different classifiers ($k = 1 \ldots K$). In order to exploit the complementary characteristics of available classifiers, the statistical combination methods described in the following sections can be used [12].

3.1 Combination by Voting Principle

Let us assume that each classifier contained in a given ensemble performs a "hard" classification assigning each input pattern to one of the M data classes. A simple method to combine results provided by these hard classifiers is to interpret each classification result as a "vote" for one of the M data classes. Consequently, the data class that receives a number of votes higher than a prefixed threshold is taken as the "final" classification. Typically, the threshold is half of the number of the considered classifiers ("majority rule"). More conservative rules can be adopted (e.g., the "unison" rule).

3.2 Combination by Bayesian Average

It is well known that some classification algorithm are able to provide an estimation of the posterior probability that an input pattern $\underline{X}$ belongs to the data class ω_i:

$$P(\underline{X} \in \omega_i / \underline{X}), \ i = 1 \ldots M \tag{3.1}$$

For example, estimates of the post-probabilities are provided by MLPs neural networks. It is straightforward to compute post-probabilities for the KNN classifier. For this kind of classifier, a natural way of combining the estimates provided by "K" different classifiers is to use the following average value:

$$P_{av}(\underline{X} \in \omega_i / \underline{X}) = \frac{1}{K} \sum_{k=1}^{K} P_k(\underline{X} \in \omega_i / \underline{X}) \tag{3.2}$$

The final classification is taken according to the Bayesian criterion, that is, the input pattern $\underline{X}$ is assigned to the data class for which $P_{av}(\underline{X} \in \omega_i / \underline{X})$ is maximum.

3.3 Combination by Belief Functions

This method exploits the prior knowledge available on each classifier. In particular, the knowledge on the "errors" made by each classifier is exploited. Such prior knowledge is contained in the so called "confusion matrix". For the z_{th} classifier C_z, it is quite simple to see that the confusion matrix can provide estimates of the following probabilities:

$$P(\underline{X} \in \omega_i / C_z(\underline{X}) = j), \ i = 1 \ldots M, \ j = 1 \ldots M, \ z = 1 \ldots K \tag{3.3}$$

where $C_z(\underline{X}) = j$ means that the classifier C_z is assigning the input pattern $\underline{X}$ to the class j.

On the basis of the above probabilities, the combination can be carried out by the following "belief" functions:

$$bel(i) = \eta \prod_{k=1}^{K} P(\underline{X} \in \omega_i / C_k(\underline{X}) = j_k), \ i = \ldots M \tag{3.4}$$

The final classification is taken by assigning the input pattern $\underline{X}$ to the data class for which $bel(i)$ is maximum.

4. Experimental Results

4.1 Selected Data Set

The selected data set consists of a set of multi-sensor remote-sensing images related to an agricultural area near the village of Feltwell (UK) [4]. The images (each of 250×350 pixels) were acquired by two imaging sensors installed on an airplane: a multi-band optical sensor (an Airborne Thematic Mapper sensor with eleven bands) and a multi-channel radar sensor (a Synthetic Aperture Radar with twelve channels related to three bands, with four polarisations for each band). For our experiments, six bands of the optical sensors and nine channels of the radar sensor were selected. As the image classification process was carried out on a "pixel basis", each pixel was characterised by a fifteen-element "feature vector" containing the brightness values in the six optical bands and over the nine radar channels considered. For our experiments, we selected 10944 pixels belonging to five agricultural classes (i.e., sugar beets, stubble, bare soil, potatoes, carrots) and subdivided them into a training set (5124 pixels) and a test set (5820 pixels).

4.2 Results and Comparisons

Our experiments were mainly aimed to investigate the following aspects:

(a) to point out that a valid alternative to the development of new classification algorithms can be constituted by the exploitation of the complementary characteristics of different classifiers by the above combination methods;

(b) to prove that the use of ensembles consisting of different classifiers allows one to obtain satisfactory classification accuracies with short designing phases.

First of all, in order to create a large "library" of classification algorithms that would allow us to build up many different ensembles, we applied various classification algorithms to the selected data set. In particular, two statistical classifiers: the Gaussian Classifier, and the k-nearest-neighbour classifier (KNN), and three neural networks classifiers: the Multilayer Perceptron neural network (MLP), the Radial Basis Functions neural network (RBF), and the Probabilistic neural network (PNN). For each classifier, a careful designing phase was carried out in order to assess the best performances provided by "single" classifiers after long designing phases. For the k-nearest-neighbour classifier, we carried out different trials with "k" values ranging from 1 up to 91. For the Multilayer perceptron Neural Networks, 5 different architectures with one or two hidden layers (15-30-5, 15-8-5, 15-15-5, 15-30-15-5, 15-7-7-5) were experimented. For each architecture, 20 trials with different random initial-weights ("multi-start" learning strategy) were carried out. For the Radial-Basis-Functions Neural Networks, different trials of the clustering algorithms ("k means") used to define the network architecture were performed (values of the number of clusters ranging from 10 up to 30 were used). As it is known, the Gaussian classifier and the Probabilistic Neural Networks do not need designing phases.

Consequently, 182 classifiers were trained and tested on the selected data set. The performances provided by the above different classifiers on the test set are summarised in table 4.1.

Table 4.1. Performances provided by the different classifiers on the test set

CLASSIFIER	LOWER ACCURACY	MEAN ACCURACY	HIGHER ACCURACY	No. of "trained" classifiers
Gaussian	79.37%	79.37%	79.37%	1
KNN	86.63%	88.36%	90.10%	46
MLP	73.45%	81.60%	89.75%	100
RBF	71.40%	78.95%	86.51%	34
PNN	88.66%	88.66%	88.66%	1

It is worth noticing that the best classifier obtained after the above "very long" designing phase is the k-nearest-neighbour ($k = 21$) with 90.10% of classification accuracy.

In order to prove that the combination of different classifiers allows one to obtain satisfactory classification accuracies with "reduced" designing phases, we combined the k-nearest-neighbour classifier using the "standard" k value ($k = 71$); this value is equal to the square root of the number of training pixels) [11], with the multilayer perceptron with architecture 15-15-5 (just one random-weight trial), and the Probabilistic Neural Network. The three classifiers provided the following classification accuracies: 88.47% for the KNN, 86.15% for the MLP, and 88.62% for the PNN. The Combination by Majority Rule of the three above classifiers provided a classification accuracy of 90.37%. The Combination by Belief Functions provided a classification accuracy of 91.29%. This experiment proves that the combination of just three classifiers performs better than the best classifier among 182. It is worth noticing that the designing phase necessary to produce these three classifiers needs of training and testing just three classifiers and allows one to obtain performances that are better than the ones provided by the best classifier (KNN, $k = 21$) obtained after a designing phase involving 182 classifiers.

Other similar experiments, that we do not report for the sake of brevity, were carried out using different ensembles (containing just neural classifiers or a "mixture" of statistical and neural classifiers). Also these experiments confirmed the conclusion that the combination of different classification algorithms allows one to obtain satisfactory classification accuracies with reduced designing phases.

5. Conclusions

We think that the experimental investigation reported in this paper shows the potentialities of the use of combination methods for integrating classification results provided by existing algorithms. In the remote-sensing field, this approach can constitute a valid alternative to the development of new classification algorithms more complex than the present ones. In particular, it seems to us that the reported results point out that the use of classifiers ensembles allows one to obtain satisfactory performances with very short designing phases.

References

1. J. A. Benediktsson, P. H. Swain, and O. K. Ersoy, "Neural network approaches versus statistical methods in classification of multi-source remote-sensing data", *IEEE Transactions on Geoscience and Remote Sensing*, vol. 28, no. 4, pp. 540–552, 1990.
2. F. Roli, S. B. Serpico, and G. Vernazza, "Neural Networks for Classification of Remotely Sensed Imagesi", *Fuzzy Logic and Neural Network Handbook*, Part 2, Chapter 15, McGraw-Hill Series on Computer Eng., C. H. Chen Editor, pp. 15.1–15.28, 1996.
3. L. Bruzzone, C. Conese, F. Maselli, and F. Roli, "Multisource classification of complex rural areas by statistical and neural-network approaches", *Photogrammetric Engineering and Remote Sensing*, 1997, in press.
4. S. B. Serpico, and F. Roli, "Classification of multi-sensor remote-sensing images by structured neural networks", *IEEE Transactions on Geoscience and Remote Sensing*, vol. 33, no. 3, pp. 562–578, 1995.
5. I. Kanellopoulos, G. G. Wilkinson, and J. Mégier, "Integration of neural network and statistical image classification for land cover mapping", in *Proceedings of the International Geoscience and Remote Sensing Symposium, (IGARSS 93), Tokyo*, 18–21 August 1993, vol. II, pp. 511–513.
6. S. E. Decatur, "Applications of neural networks to terrain classification", in *Proceedings International Joint Conference on Neural Networks 89, Washington D.C.*, vol. 1, pp. 283–288, 1989.

7. J. Lee, R. C. Weger, S. K. Sengupta, and R. M. Welch, "A neural network approach to cloud classification", *IEEE Transactions on Geoscience and Remote Sensing*, vol. 28, no. 5, pp. 846–855 1991.

8. H. Bischof, W. Schneider, and A. J. Pinz, "Multispectral classification of Landsat-images using neural networks", *IEEE Transactions on Geoscience and Remote Sensing*, vol. 30, no. 3, pp. 482–490, 1992.

9. M. R. Azimi-Sadjadi, S. Ghaloum, and R. Zoughi, "Terrain Classification in Sar Images Using Principal Components Analysis and Neural Networks", *IEEE Transactions on Geoscience and Remote Sensing*, vol. 31, no. 2, pp. 511–515, 1993.

10. Y. Salu, and J. Tilton, "Classification of multispectral image data by the Binary Diamond neural network and by non-parametric, pixel-by-pixel methods", *IEEE Transactions on Geoscience and Remote Sensing*, vol. 31, no. 3, pp. 606–617, 1993.

11. K. Fukunaga, *Introduction to Statistical Pattern Recognition*, Academic Press, Inc., New York, 2nd edition, 1990.

12. L. Xu, A. Krzyzak, and C. Y. Suen, "Methods for combining multiple classifiers and their applications to handwriting recognition", *IEEE Transactions on Systems, Man, and Cybernetics*, vol. 22, no. 3, pp. 418–435, 1992.

Integrating the Alisa Classifier with Knowledge-Based Methods for Cadastral-Map Interpretation

Eleni Stroulia and Rudolf Kober

Research Center for Applied Knowledge Processing (FAW)
Helmholtzstr. 16, 89081 Ulm, Germany.
e-mail: {stroulia, kober}@faw.uni-ulm.de

Summary. Alisa is a learning, statistical, texture classifier for single- and multi-class classification. Its process is based on the examination of images using a set of universal (i.e., independent of the domain of the images under examination) features. Given a set of pre-classified examples, it computes a subset of these features for a small window (i.e., the *analysis token*) centred at each image pixel, and creates a histogram of occurrences of the distinct feature values in the training data. After training is completed, given an unknown image Alisa examines this in the same way, and generates an isomorphic image, each pixel of which represents the normality of the corresponding pixel in the input image (or, in multi-class classification, the class in which it belongs). In this paper, we discuss the integration of Alisa with knowledge-based methods for recognising line thickness in cadastral maps.

1. The Alisa Classifier

Based on the paradigm of Collective Learning Systems (CLS) theory [2], Alisa (Adaptive Learning In Signal Analysis) was originally designed to detect anomalies in images. Alisa learns the characteristics of a class of images from examples of the class using statistical methods. Then, it is able to detect faults in other images of the class. The original Alisa classifier was extended in order to be able to process single- or two-dimensional signals, to classify signals from different sensors separately or to fuse them (when they are isomorphic) and classify their combination, and to address multi-class classification tasks.

Alisa is a three-layer, hierarchical network of non-learning and learning cells arranged in parallel channels. The network consists of the analysis, hypothesis and synthesis layers.

The Analysis Layer. In the analysis layer, the input image may first be processed (e.g., spatially filtered) to produce a transformed image. One potentially useful transformation could be low-pass filtering with different cut-off frequencies resulting in the analysis of the image at different resolutions: as the cut-off frequency decreases the details disappear, until only the large structures are visible. An alternative transformation is the orthogonal decomposition of the input image in different frequency channels using the wavelet transformation, or the "mean image" or the "variance image" of the original input.

Following such a transformation, the image is re-represented by feature vectors, calculated for a small analysis token centred at every pixel. The feature vectors are constrained to 16 bit, allowing for a maximum of $2^{16} = 65536$ distinguishable feature vectors. The size of the analysis token, the features used for the analysis and re-representation of the image, and their quantisation are input parameters to Alisa. Currently available features include gradient direction, gradient magnitude, pixel mean, pixel activity, gradient direction activity, and contrast. The useful dynamic range of each selected feature value is then quantised to the pre-specified number of bits. Values outside the feature's dynamic range are placed in either an overflow or underflow category, to maximise the resolution over the most indicative ranges of the feature values. The resulting values are concatenated to produce the 16-bit feature vector.

The Hypothesis Layer. In the hypothesis layer, statistical knowledge about the occurrences of the different feature vectors in the training images of the different classes is used to produce the channel hypotheses. The statistical knowledge is stored in the *state transition matrix* (STM), built up for each channel and for each class, during the training phase by counting the number of occurrences of each distinct feature vector in the training images. The STM is essentially a class-conditioned histogram of the feature vectors of the training images, for each different channel and for each different class. A channel hypothesis is an image, isomorphic to the original one. In anomaly-detection classification tasks, the value of each pixel in the channel hypothesis indicates the normality of the corresponding pixel value in the original image. In the case of multi-class classification tasks, the value of each pixel in the channel hypothesis indicates the class to which the corresponding pixel in the original image most probably belongs.

The Synthesis Layer. In the synthesis layer, the channel hypotheses are combined into a single super hypothesis, also represented by a image isomorphic with the original one. Similar to the channel hypotheses, each pixel in the super hypothesis represents, in anomaly-detection classification tasks, the normality of the corresponding pixel in the image under investigation, and, in the case of multi-class classification tasks, the index of the class in which this pixel belongs.

In anomaly detection, the value of each pixel in the synthesised super hypothesis is computed by averaging all the non-normal values of the corresponding pixels in all the channel hypotheses. In multi-class classification, the value of each pixel in the super hypothesis is computed as the index of the "most-probable class", where the probability of each class is computed by a weighted sum of the probability of this class in each channel hypothesis.

1.1 Image Classification with Alisa

The image-classification process in Alisa consists of three phases: the training phase, in which the texture characteristics of images of different classes are learned; the control phase, in which unique images of the training class are classified to provide a statistical basis for subsequent tests; and the test phase, in which arbitrary images are classified.

The Training Phase. During the training phase, Alisa counts the occurrences of each possible feature vector, and creates the STMs, i.e., the histograms of the class-conditioned probabilities of the feature vectors in the training data.

The Control Phase. The control phase enables the statistical verification that a sufficient number of training images have been sampled, and that the mean and variance of the training class has stabilised. In this phase, learning, i.e., the updating of the STMs, is disabled, and Alisa is presented with a unique sample of images of the training class, i.e., images not presented during training. Satisfactory classification of these images suggests that training has been completed.

The Test Phase. The test phase operates in the same way as the control phase, except that the class of each input image is arbitrary. For example, if in anomaly-detection a test image contains only structures which are characteristic of the training class, then Alisa will evaluate it as entirely normal; its channel and super hypotheses will be similar to those of the control images. If the test image contains anomalies, or belongs in an entirely different class than the training images, the hypotheses will reflect this with consistently low normality values. The isomorphism of the hypotheses with the original image allows the spatial localisation of any anomalies. The value of a non-normal pixel in a given channel hypothesis provides a quantified degree of normality at that channel for the corresponding pixel in the original image, while the super hypothesis provides a summarised result of each pixel for all channels.

2. Cadastral Map Interpretation

Cadastral maps (plans) are drawings of city plans showing building lots, houses, and utility lines. The problem of extracting information from digitised, hand-drawn cadastral maps is a complex one, [1, 4], an important subproblem of which is the recognition of the different types of lines that appear in them in terms of their thickness. Since, in most cases, different types of objects (such as building blocks, houses, utility lines etc.) are drawn with lines of different thickness, recognising such line thicknesses in the map are crucial steps in recognising these objects.

To address these problems we employed Alisa, for the recognition of the different types of line thickness. Alisa's training enables it to learn the characteristic features of an image class given examples of this class, thus making it quite robust with respect to local image anomalies. That is the main reason why we chose it to recognise the different line thicknesses in a cadastral map.

Another reason was that another important source of information in some types of such maps is the texture of the different patterns used to denote closed-shape areas, the recognition of which would also be a natural task for Alisa.

In spite of Alisa's robustness with local anomalies in the image, the poor quality of the original maps and noise introduced while scanning them results sometimes in misclassifications of line segments. To address this problem we have introduced some a-priori domain knowledge in the overall process. In several cases, the intersections of particular types of lines are annotated with special symbols, such as circles of a particular radius for example. In such cases, we use classic pattern-recognition methods to recognise these special symbols, and then use the knowledge of where these special symbols exist in the image to provide additional constraints in classifying the line segments that intersect at them.

To summarise, to build a system for recognising the different types of lines in cadastral maps, we integrated several different types of methods: we use (a) classic image-processing methods to identify the different line segments, (b) pattern-matching methods to identify particular symbols defining line boundaries, (c) the Alisa classifier to categorise the different line segments, and (d) model-based methods to synthesise the information produced by all of the above.

2.1 The Functional Architecture

In this section, we briefly discuss the roles that these methods play in the overall-system's processing as shown in figure 2.1.

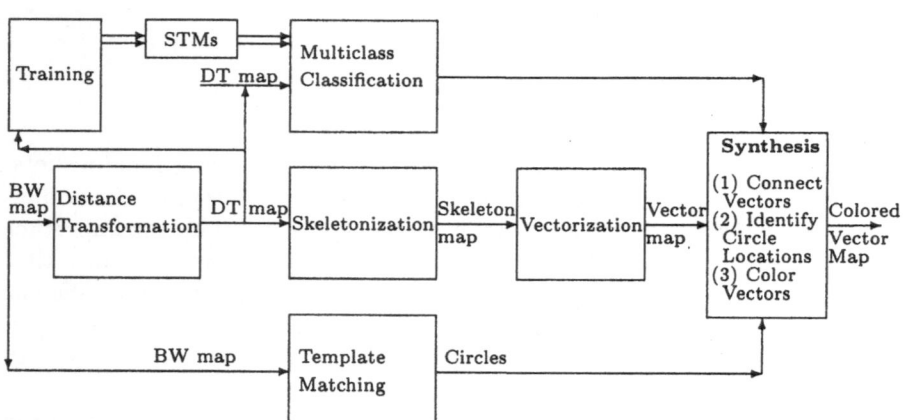

Fig. 2.1. The functional architecture of the application.

Distance Transformation. This step takes as input a black-and-white cadastral map and produces as output a gray scale map where each pixel in each foreground structure has as value its distance from the background [3]. The processing begins at the top-left corner of the image and ends at the bottom-right corner, colouring each pixel with its distance from the background. The distance of an external pixel is 2 if a pixel directly above or below it, or a pixel directly left or right of it belongs to the background, and 3 if a neighbouring pixel in the diagonal belongs to the background. The distance of each inner pixel is computed as the minimum between the distances of the pixels above and to the left, plus two. This pass ignores the information at the right of and below a pixel. Thus a second pass, beginning from the bottom-right corner towards the top-left one is necessary. In this pass, the distance of each inner pixel is computed as the minimum between its current value and the values of the pixel below it and the pixel on its right, plus two.

Skeletonisation. This step takes as input a distance-transformed map and produces as output a map where all foreground structures are "thinned" down to one pixel along its dominant direction. The skeletonisation step recursively removes all external pixels, until only the pixels with the highest colour value remain. A pixel of a lower colour value is not removed, when the only way between two neighbouring pixels includes the pixel under investigation. Each step of the process (i.e., removal of pixels of colour 2, 3 and so on) is executed beginning at each corner of the image towards the opposite corner, thus, making sure that the resulting skeleton of the foreground structures is symmetric.

Classification of the Skeleton-Pixels Topology. This step classifies the pixels pf the skeleton, according to their topology, in (a) isolated nodes (pixels without neighbours), (b) end-nodes (pixels with one neighbour), (c) intermediate nodes (pixels with two neighbours) and (d) star-nodes (pixels with more than two neighbours). Based on this information, the topological structure of the skeleton, i.e., its nodes and its edges, is produced.

Vectorisation. The goal of the vectorisation step is the transformation of the skeleton into approximate geometric structures. This step takes as input a skeletonised map and produces as output a list of the different line segments which it contains. For each edge identified in the previous step, the minimum number of line segments that can describe it are identified. An edge can be described by a single line segment, if the distance of the median pixel of the segment defined by its two end-nodes from the edge is smaller than the average edge thickness. If this is not the case, the process recursively proceeds to describe the two halves of the edge as defined by its two end-nodes and its median pixel.

Symbol Identification by Pattern Matching. This step takes as input the original black-and-white map and produces as output an image containing the positions of particular symbols in this map. To effectively recognise a symbol, a mask is needed such that it generates a maximum response when and only when it is convoluted with an area in the image where the pattern occurs. This mask is used as a non-linear filter to transform the original black-and-white image. The result of the filtering process is a map where the locations of similar circle-like patterns in the image are white and the rest of the image is black. With a subsequent high-pass filter, only these locations with a high enough response remain white, thus eliminating spurious circle-like patterns for the set of cornerstone circles. In our system, we used this method to identify the positions of the building-lot cornerstones.

Alisa Classification. This step may take as input the input black-and-white image, or the distance-transformed one, and produces as output a set of channel hypotheses; each channel hypothesis represents, as we discussed above, the class membership of each pixel in the foreground structure of the map.

Knowledge-Based Synthesis. After having vectorised the input image, having classified each pixel of the image with respect to the thickness of the line in which it belongs, and having filtered the original image with the patterns of any special symbols that might denote line boundaries, all of this information is fused into a single output. This final step takes as input all the above information, and produces as output a vectorised image where each vector is "coloured" with the class in which the thickness of the line it represents belongs. This synthesis task involves three different substeps.

(1) **Connection of Collinear Vectors** Often, the vectorisation process results in overly fragmented vectors. For example, due to noise and local thickness anomalies a single straight line may be split into two or more vectors. The goal of this task is to reconnect such artificially separated vectors. To that end, it identifies pairs of vectors which have the same direction and one begins where the other ends, and substitutes them with a single vector.

(2) **Identification of Special Symbol Locations** The filtering process of section 2.1 results in a black-and-white image where white pixels belong in structures in the original image resembling the pattern. This task, given this image, identifies the locations of these formations, and collects them in a list.

(3) **Vector colourisation** The final step in the synthesis task is the actual colourisation of the image vectors. For each vector, this step identifies the class which appears most often among its pixels in the channel hypotheses produced by Alisa. This class will constitute the colour class for the vector at hand. Notice, that in this particular Alisa application there is no need for the synthesis of the channel hypotheses into a super hypothesis at the pixel level. The synthesis is done instead at the vector level. If, however, the vector lies on the line connecting two special symbols, then the class of the vector is determined by high-level knowledge regarding the type of lines that lie within these symbols. Figure 2.2 shows an original map, the map after the classification of its pixels by Alisa, and finally, the map after the knowledge-based synthesis that has classified its vectors. We can see how the noise has

Fig. 2.2. (a) The input map (left), (b) the map classified by ALISA (top right), and (c) the map after the knowledge-based synthesis (bottom right)

decreased especially in the closed shapes representing houses, as well as in the dashed line below the top-left–bottom-right diagonal.

3. Evaluation

We have performed two different experiments: in the first, we used Alisa to classify the original digitised black-and-white maps, and in the second, we used it to classify the same maps after they were distance-transformed (i.e., after their pixels were given a gray value according to their distance from the background). As there is a variety of line types whose thickness may range from 2-3 to 9-10 pixels wide, in all these experiments we had to use Alisa with four or five different analysis tokens. Alisa produced much better results with distance-transformed maps, and this was the type used for the experiments below.

Then we tested the system with a set of artificial maps, and two different types of actual maps. We had 95-97% recognition rate with the artificial maps (the recognition rate is calculated as the percentage of the correctly classified vectors in the image). With the actual maps this percentage reduced to 92% in the first type and 89% in the second type. With the first type of maps, we tested the system with two different map subtypes, with more and less symbolic information written on them. This difference did not affect the classification rate. This classification rate will not suffer considerably, as long as there is at least 2 pixels difference in the thickness of "neighbouring" classes, and the map is not too clattered.

4. Summary

In this paper we described a prototype system for recognising and classifying according to their thickness lines in cadastral maps. Because there exist several different types of structures in such maps, we integrated different types of methods in this prototype:

(1) To recognise regular structures, we used pattern-recognition techniques. For each regular structure, we designed a filter that generates a maximum response when convoluted with an area which includes such a structure.

(2) To recognise line segments, we used a "standard" image-processing method, which includes distance-transformation, skeletonisation and vectorisation steps.

(3) To classify different line segments according to their thickness, we used the Alisa classifier which learns from a set of examples and robustly classifies structures according to their texture.

(4) Finally, we used knowledge-based techniques to integrate the information from one method in the process of another. For example, we used the

vector information in the synthesis step of Alisa, and we even ignored the Alisa classification when it conflicted with the suggestions of domain rules applied to the recognised regular patterns.

Acknowledgement. The authors would like to thank Christian Flick for his work on the development of the classic image-processing methods employed in the prototype system described here, and also Christian Schiekel and Rainer Ossig for their helpful insights on Alisa.

References

1. Boatto, L., Consorti, V., DelBuono, M., Di Zenzo, S., Eramo, V., Esposito, A., Melcarne, F., Meucci, M., Morelli, A., Mosciatti, M., Scaric, S., Tucci, M., "An interpretation system for land and register maps", *IEEE COMPUTER*, July 1992.
2. Bock, P., Klinnert, R., Kober, R., Rovner, R., and Schmidt, H., "Gray Scale ALIAS", *IEEE Special Transactions on Knowledge and Data Engineering*, March, 1992.
3. Lange, M., "Segmentierung von Konturen auf der Basis von Kruemmungs-berechnungen", *IEEE Pattern Analysis and Machine Intelligence*, 1993.
4. Maderlechner, G., Mayer, H., "Conversion of high-level Information from Scanned Maps into Geographic Information Systems". In the *Proceedings of 3rd International Conference on Document Analysis and Recognition (ICDAR) 1995*.

A Hybrid Method for Preprocessing and Classification of SPOT Images

Shan Yu and Konrad Weigl

INRIA, 2004 route des Lucioles,
B.P. 93 - 06902 Sophia-Antipolis Cedex, France
e-mail: {yu,weigl}@sophia.inria.fr

Summary. In this paper, we present a hybrid method for preprocessing and classification of satellite images. The preprocessing consists of computing texture measures of the images and initialising the probabilities of pixels belonging to different land-cover classes. The objective of the preprocessing is twofold: increasing discrimination power and removing irrelevant characteristics. The classification process consists of assigning a class to each pixel, with a special interest in detecting urban areas as completely as possible with the aid of a priori knowledge. This interest stems from the possible requirement of detecting urban areas on satellite images (even small villages in the countryside) while ignoring some classes (such as parks) in cities. We shall show how this requirement is translated into constraints imposed in our classification process. Experimental results are illustrated through a SPOT image containing a coastal town.

1. Introduction

We consider a real world problem — land cover classification using remotely sensed images. More specifically, we look for a mapping from the pixel space to the class space. This relationship, in the real world, is not necessarily deterministic in itself, due to incomplete information, class mixtures on the ground, lack of spatial resolution or spectral bandwidth, etc. Conventional image processing methods are not sufficient to solve this problem. A priori knowledge about the world model is often needed in order to improve the classification results. Nevertheless, perfect classification is hardly feasible, nor necessarily wanted. Very often, people are content with solutions which meet a set of specific requirements.

We are interested in dealing with some typical requirements that users might ask, but not usually expressed and satisfied by current land cover classification scheme. For instance, one of the typical requirements is to identify as many patches of class A in an image as possible, while ignoring patches of classes B, C, D in patches of class A. An example of this is to detect urban areas, ignoring parks in cities. Such a requirement may prohibit the use of some preprocessing operators in order to preserve the data concerning class A (or urban area in this paper). It is obvious that not all purposes can be fulfilled by the same approach, thus a compositional approach, or a modulated method is useful: build a specific sequence of processing for each purpose.

Our hybrid method consists of two parts: preprocessing of image data and classification. The purpose of preprocessing is to remove irrelevant aspects such as those due to variances of rotation and translation of textures, and extract relevant features to increase the discrimination power between different classes. This step is crucial for a successful classification. For this purpose, we first compute the texture measures of the image. Then we look for the relationship between non-linear combinations of feature values and the corresponding class-membership of a pixel. This is performed by a feed-forward neural network, trained by a novel rapidly converging technique, referred to as the *projection learning algorithm* (PLA) [5, 6]. This very fast algorithm gives the initial probabilities of pixels belonging to different classes. The advantage of using such a neural network is that its performance is independent of the statistical distribution of image data.

The classification process uses a priori knowledge to favour urban area detection. Two kinds of a priori knowledge are available: common sense knowledge about the world model and map knowledge about the urban area of the scene. We use Markov random field (MRF) modelling of images to introduce such a priori knowledge, as well as constraints on the classification results among neighbouring pixels. The basic idea of the approach is to repeatedly apply an optimisation algorithm [7] to label the image, compare the (intermediate) labelling result with the map information, and modify the potentials in the energy function so that the labels become closer to the expected values.

In what follows, we shall describe each part of our method, and illustrate experimental results. Finally, we conclude with discussions of future research directions.

2. Preprocessing of Image Data

2.1 Texture Measurement

Urban scene satellite images are often highly textured. Therefore we use texture measures as a homogeneous criterion to discriminate different classes of land cover. After having tested different texture measures, we select the following four, implemented by Gertner and Andre [2]:

- The first measure is the density of the norm of the image's Laplacian. The Laplacian is computed in the classical manner, i.e. the curvature operator was implemented in a discrete way both in the x and y directions. Then the modulus of the two values is taken. The assumption is that landscapes would either have a very small (within fields) or very large (boundaries of fields) modulus, and cities would be characterised by a medium-valued modulus.
- The second is the orientation of the gradient of the image. The gradient is computed using the classical Canny-Deriche operator [1]. A window of 15×15 pixels is used to compute the local average (vector-valued). After that, we compute the scalar distance between the vector-valued gradient

of a pixel and the local average of the gradient. This operator is a measure of the local directional variance of image gradient, which is expected to be large in an urban area and small elsewhere.

– The last two measures derive from the local histogram: the phase and the amplitude. The phase allows us to differentiate between dark-coloured regions such as water bodies and light regions such as cities. The amplitude can be interpreted as the distance to a fully noisy signal. It will thus be relatively large for rather uniform regions such as fields, and small for regions with many different grey-scale values such as towns.

These texture measures, though not fully discriminating, remove the majority of irrelevant variabilities. Each of these measures are then used as a component of the feature vector of the image, fed as input data to the neural network for initialisation of class-membership of the image.

2.2 Initialisation

Here a feed-forward neural network is used. Such networks are usually taught by back-propagation algorithms, which are easy to use but rather slow to converge. Though there have been a multitude of successful attempts to improve the speed of the algorithms, the velocity is still quite low.

We consider the problem from another point of view: we view neural networks with multi-layer perceptrons as non-orthogonal bases in a function space whose dimensions correspond to the number of learning samples. Suppose that we have a neural network with one input neuron, two hidden neurons, and one output neuron. Suppose also that we have only three samples x_0, x_1, x_2 for function approximation. Noted by g_1, g_2 the basis functions computed by the two hidden neurons, and A_1, A_2 the weights from the hidden neurons to the output neuron. We can thus represent these values as vectors in a three-dimensional function space (cf. figure 2.1). The basis functions span a sub-manifold in this space. The function F to be approximated is projected onto this sub-manifold, noted by A.

The objective of learning in such a network is thus to minimise the distance between the function to be approximated F and its projection on the sub-manifold A. This is achieved by dynamically rotating and shifting the base through modification of the basis functions' parameters of the network. We call this learning approach the *Projection Learning Approach* (PLA). If the network has learnt and generalised well, then the approximated function will be close to the function itself in other dimensions and for other inputs not in the training set.

Let D be the distance (usually a mean-squared error) to be minimised, and γ the vector of all the input to hidden-layer weights:

$$D(F, A(\gamma)) = \sum_x (F(x) - A(\gamma, x))^2 = \sum_x (F(x) - \sum_i A^i g_i(x))^2$$

Approximated function $\vec{A} = A^1\,\vec{g}_1 + A^2\,\vec{g}_2$

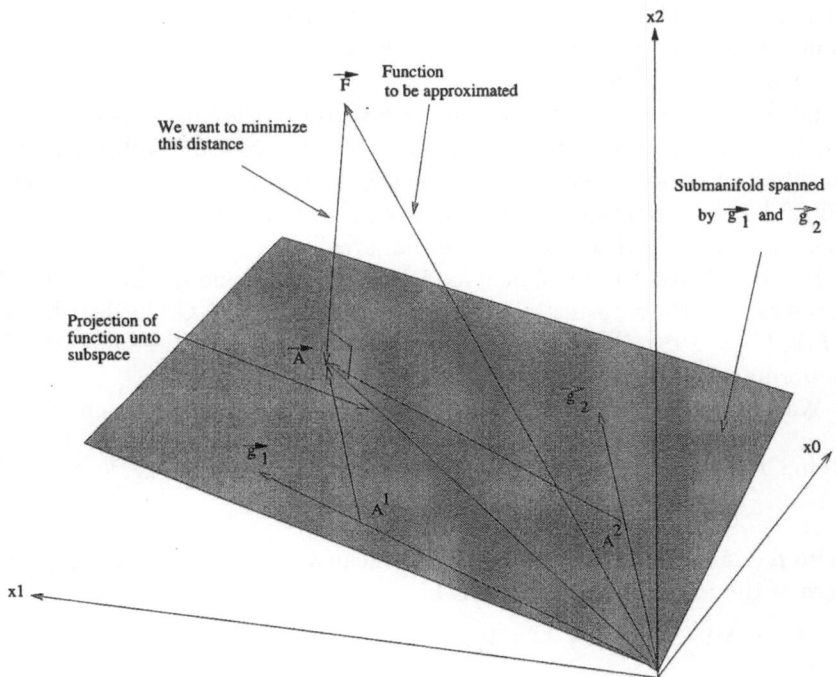

Fig. 2.1. A representation in a three-dimensional function space

Some computation leads then to the final formula

$$\frac{dD(F, A(\gamma))}{d\gamma} = -2\sum_{x}\left(F - \sum_{i} A^i g_i(x)\right)\sum_{i} A^i \frac{dg_i(x)}{d\gamma}$$

where $i \in (1, \cdots, N)$, N is the number of hidden-layer neurons. The solution can be obtained by using metric tensors or some other methods, such as gradient descent methods. Refer to [5, 6] for details.

3. Classification with the Aid of A Priori Knowledge

The objective of this part is to classify the image sites, with a special interest in detecting urban areas as completely as possible. We use Markov random field modelling of images to introduce knowledge about the urban area extracted from the corresponding geographical maps and constraints on the classification results among neighbouring pixels.

Let $S = \{s_1, s_2, \cdots, s_n\}$ be the set of image sites (the same set of sites will also used to index the map image). A global discrete labelling L assigns to each site i one label $\lambda_i \in \Lambda_i$, $\Lambda_i \subseteq \Lambda = \{\lambda_1, \lambda_2, \cdots, \lambda_m\}$, $i = 1, 2, \cdots, n$. Let c denote a clique, and $\mathcal{C}$ the set of all cliques. The number of sites in clique c is referred to as its degree : $\deg(c)$. Let L_c be the restriction of L on the sites of a given clique c.

Let $Y = \{y_i, 1 \leq i \leq n\}$ be the feature vector of the image. Let the set of sites be partitioned into two subsets: $S = S' \cup S''$, $S' \cap S'' = \emptyset$, where S' is the subset of sites corresponding to the urban area in the map image. Let L' and L'' be the restriction of labelling L on S' and S'', respectively. Thus all labels in L' are equal to the label "urban area".

By using Markov random field modelling of images and the Bayes formula, we can express the a posteriori probability $P(L|Y, L')$ as a Gibbs distribution: $P(L|Y, L') = \frac{1}{Z} \exp\left(\sum_{c \in \mathcal{C}}(-V_{cL_c})\right)$, where V_{cL_c} is a potential function, Z is the normalising constant.

We consider clique potentials of orders 1 and 2 in the above function:

$$U_1(L, Y) = \sum_{c \in \mathcal{C}, \deg(c)=1} V_{cL_c} = \sum_{s \in S} \left(\log|\Sigma_{\lambda_k}| + (y_s - \mu_{\lambda_k})^T \Sigma_{\lambda_k}^{-1}(y_s - \mu_{\lambda_k})\right),$$

(3.1)

where μ_{λ_k} and Σ_{λ_k} are respectively the mean value and the covariance matrices of the feature vector for class k.

$$
\begin{aligned}
U_2(L, Y) &= \sum_{c \in \mathcal{C}, \deg(c)=2} V_{cL_c} \\
&= \sum_{(i,j) \in C} P_{ij}(\lambda_i, \lambda_j) \\
&= \frac{1}{k^2} \sum_{s \in R_k(i)} P_s(\lambda_s) \cdot P_{s+\delta}(\lambda_{s+\delta}), \quad j = i + \delta, \quad \delta \in \Delta. (3.2)
\end{aligned}
$$

where $P_s(\lambda_s)$ is the initial probability of site s having label λ_s. $R_k(i)$ represents a $k \times k$ window centred at site i. δ denotes the neighbourhood relation between i and j.

We use a labelling algorithm [7] to assign a label to each image site. The key idea of the present approach is follows: we compare the labelling result with the map information, modify the potentials in the energy function, and re-apply the labelling algorithm. This process is repeated so that the labels are closer to the expected values.

We assume that the growth ratio of the urban area, noted as τ, after the map was produced is available. This ratio can be obtained, for example, from estimating population growth or from town planning projects [4]. Let α be the growth ratio of the urban area computed by comparing the (intermediate) labelling result and the map information. Let λ_0 be the "urban area" label, d the the number of neighbours of a site, and v_i the number of neighbours of site s_i having label λ_0. Let γ be an adaptation parameter, $0 < \gamma < 1$.

If $\alpha < \tau$, the detected urban area is smaller than the expected one, thus:

$$P_i(\lambda_0) = P_i(\lambda_0) + (1 - P_i(\lambda_0)) \cdot \gamma \cdot \frac{v_i}{d};$$

$$P_i(\lambda) = P_i(\lambda) - P_i(\lambda) \cdot \gamma \cdot \frac{v_i}{d}, \quad \lambda \neq \lambda_0.$$

If $\alpha > \tau$, the detected urban area is larger than the expected one, thus:

$$P_i(\lambda_0) = P_i(\lambda_0) - P_i(\lambda_0) \cdot \gamma \cdot \frac{1}{1 + v_i};$$

$$P_i(\lambda) = P_i(\lambda) + \left(\frac{1}{|A| - 1} - P_i(\lambda) \right) \cdot \gamma \cdot \frac{1}{1 + v_i}, \quad \lambda \neq \lambda_0.$$

Note that the modification of the probabilities also takes into account neighbourhood information (through v_i/d). The adjusted probabilities are then taken into account in the computation of potentials of cliques of order 2 using relation (3.2).

4. Experimental Result

We illustrate the experimental result of the work on a SPOT XS3 image containing a coastal town, shown in figure 4.1-(a). Figure 4.1-(b) gives symbolic information about the urban area of the scene extracted from the corresponding geographic map. The white colour represents urban areas, the dark colour the non-urban areas. It is assumed that such information is not perfect. Figure 4.1-(c) is the classification result obtained by the present method. The white colour represents the urban areas, the grey one the countryside fields, and the dark one the water bodies. For sake of comparison, we also give in figure 4.1-(d) the classification result obtained with no use of map knowledge.

The undetected urban region at the centre of figure 4.1-(d) is due to the fact that in this region there is a museum and some monuments surrounded by green area. Hence the texture feature of this region does not resemble that of the other part of the urban area. It is detected as urban area in figure 4.1-(c) because the present method has some specific information about this area issued from the map. We also see that the canal in the town is detected in figure 4.1-(c), owing to the good initialisation of the neural network. It is misclassified as countryside in figure 4.1-(d).

5. Conclusion

We have presented a hybrid method for preprocessing and classification of satellite images. The step of preprocessing the image data is crucial for a

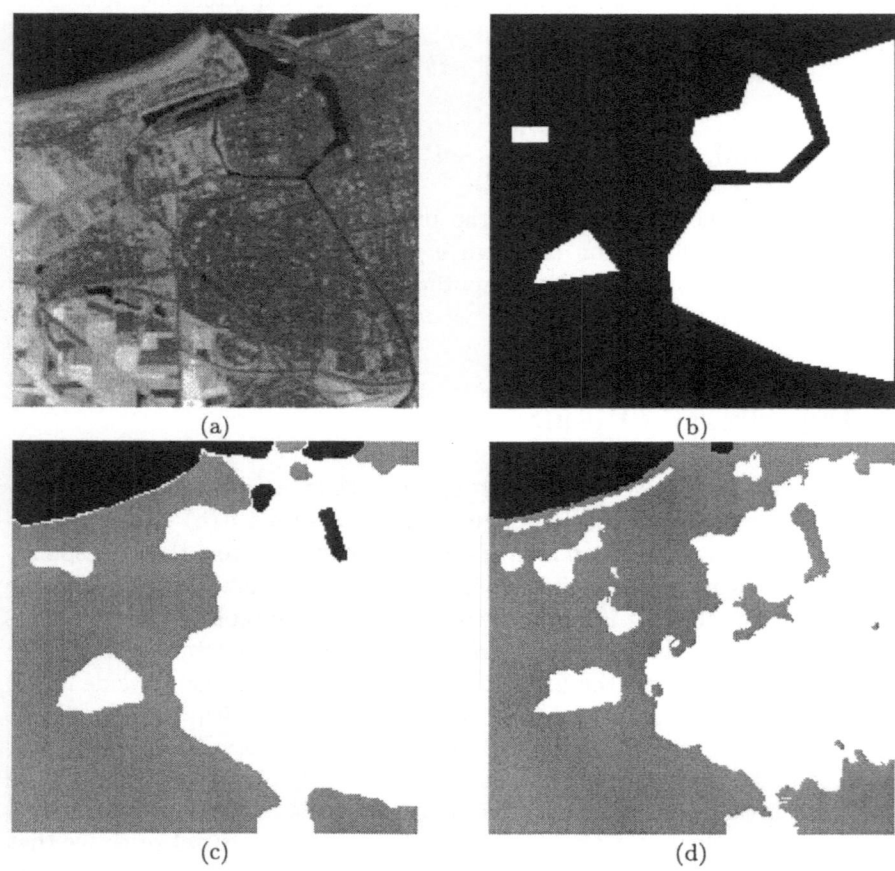

Fig. 4.1. Spot image classification using a priori knowledge. (a) A SPOT XS3 image containing a coastal town (Calais in France). (b) Symbolic information about the urban area extracted from a geographic map. (c) Classification result obtained by the present method. (d) Classification result obtained without using map information.

successful classification of the image. A priori knowledge coming from geographical maps is used in order to meet the requirement of detecting as completely as possible urban areas, while ignoring some other classes in cities. The hybrid system is inherently well suited to deal with such requirements, since their modular structure, with a corresponding maximal decoupling of functionality between the modulus, enables them to be customised to a strong extent.

The experiments with the hybrid system are ongoing. We intend to extend the investigation to meet some other requirements that users might ask. The purpose of this work is to classify satellite images, with the preoccupation of detecting as many urban areas as possible. Under this consideration, information extracted from the geographic map only concerns urban areas. The algorithm itself also favours urban area detection. Nevertheless, we could extract other kinds of information from maps or from some geographical information data base, and modify the algorithm so as to meet other requirements.

References

1. R. Deriche, "Using Canny's Criteria to Derive a Recursively Implemented Optimal Edge Detector", *International Journal of Computer Vision*, vol. 1, no. 2, pp. 167–187, 1987.
2. V. Gertner and H. Andre, "Extraction d' éléments texturés dans les images SPOT", ("Extraction of textured elements from SPOT images"), ISTAR/INRIA-Sophia Antipolis/ESSI 3, Research Report, May-July 1990, in French.
3. J. D. Paola and R. A. Schowengerdt, "A Review and Analysis of Neural Networks for Classification of Remotely Sensed Multispectral Imagery", Technical Report 93–05 of Research Institute for Advanced Computer Science, NASA Ames Research Center, June 1993.
4. G. J. Sadler and M. J. Barnsley, "Use of population density data to improve classification accuracies in remotely-sensed images of urban areas", *First European Conference on Geographical Information Systems*, pp. 968–977, April 1990, Netherlands.
5. K. Weigl and M. Berthod, " Neural Networks as Dynamical Bases in Function Space", Research Report INRIA, RR-2124, 1993.
6. K. Weigl, "The Application of Neural-Network Algorithms to Early Vision", PhD thesis, University Nice-Sophia-Antipolis, France, 1994.
7. S. Yu and M. Berthod, "A Game Strategy Approach to Relaxation Labelling", *Computer Vision and Image Understanding*, vol. 61, no. 1, pp. 32–37, 1995.

Testing some Connectionist Approaches for Thematic Mapping of Rural Areas

Leopoldo Cortez, Fernando Durão, and Vitorino Ramos

Centro de Valorização de Recursos Minerais (CVRM),
Technical University of Lisbon,
Instituto Superior Tecnico - Mining and Georesources Dept.,
Av. Rovisco Pais - 1096 Lisbon, Portugal.
e-mail: cvrm@alfa.ist.utl.pt

Summary. An overview of several supervised classification methodologies applied at the research centre CVRM to automatic classification of remote sensing data is presented. Some classification methodologies based on Artificial Neural Networks are roughly described, with emphasis on Multilayer Feed-forward Networks (MLFN) trained with the Back-propagation algorithm and on Probabilistic Neural Networks (PNN). A technique under study at the CVRM for reducing training time, conjugating Principal Components Analysis and Genetic Algorithms, is outlined. Finally a case study is presented, describing an application of a MLFN and a PNN to a pixel based multispectral (6 bands) Landsat TM data from the rural Moura-Ficalho area (South of Portugal). The results were mapped and the performances compared, by means of confusion matrices and also by the computing time used, with a typical parametric approach such as the classical Maximum Likelihood Classifier and with field work interpretation.

1. Introduction

Satellite remotely sensed data combined with Geographical Information Systems (GIS), has become a very important tool to get information about the Earth, and is more and more indispensable in numerous applications, such as land use mapping, observing and monitoring changes in the natural environment or prospecting for mineral resources. Consequently, several techniques for extracting information from remotely sensed data have been developed. These generally use multispectral observations of the same location and therefore the analysis of data, in order to divide it in a meaningful set of classes, is a multi-dimensional pattern recognition problem. The classification techniques employed, either supervised or unsupervised, must be accurate, robust and quick to apply.

In supervised classification techniques, which are the most widely used, the generalisation ability of the classifier is a crucial factor for successful employment for wide area mapping; its generalisation ability is related both to the mathematical sophistication of the classifier model and to the extent to which the ground data are representative and reliable [1].

Traditional supervised algorithms used in multi-dimensional data classification can be roughly subdivided into three main groups [2]:

- *Minimum-distance classifiers*, based on the definition of a Euclidean distance to the centroid or barycentre of each class presented in the images. A set of training samples is used to calculate the centroid coordinates (into nD space) of each one of the k classes, and each pixel is associated with the least distant class according to the value of the Euclidean distance function $d(x, k)$ defined in relation to each k class centroid.
- *Parallelepiped classifiers*, are similar to the minimum-distance classifier, in the sense that they also need a centroid definition of each class. They are intended to model the shape of each training nD cloud by defining nD parallelepiped centred at each class centroid.
- *Statistical classifiers*, that make an explicit statistical hypothesis about the sample distribution of each class. For instance, the commonly used maximum likelihood classifier assumes for each class a multi-Gaussian distribution.

All these algorithms present some drawbacks: the first do not take into account the shape of the nD clouds of training pixels; the parallelepiped and statistical classifiers do not take into consideration irregular and complex shapes of the training clouds; overlapping can occur frequently.

Due to the rather poor modelling of the shapes of the training clouds when they present irregular and complex shapes, other approaches are needed to improve the classification procedures. At the "Centro de Valorização de Recursos Minerais (CVRM)" of the Technical University of Lisbon two main alternative methodologies have been used for some years in the processing and interpretation of remote sensing images.

The first one is the use of morphological operators in several steps of supervised multidimensional classification methods. The key idea of mathematical morphology is complementary to other methods, because it consists of a geometrical approach to image processing using set operators generated by a structuring element, acting on the objects independently of their size, shape and grey level. These operators can be grouped in a set of three steps: low-level feature extraction, morphological enhancement of classification and low- level generalisation algorithms [3].

In *low-level feature extraction*, the alternating sequential filters have been defined [4] as families of openings and closings parametrised by a positive number λ (the size of the structuring element). This filter is a sequence of openings and closings acting alternatively and sequentially upon the peaks (through the openings) and the valleys (through the closings) of increasing length (from 1 to λ), suppressing the former and filling the latter of lengths $\leq \lambda$. In this way, topological features are taken into consideration and subsequent pixel-based classifiers can perform more accurately.

In *morphological enhancement*, the operators can act in several ways: for instance, in the modelling of cluster hulls of training samples, they can eliminate small unconnected points by erosion followed by a reconstruction, or define criteria for processing overlapping regions.

Finally, in the last step of the classification procedure, *low-level generalisation* algorithms can be used to perform the spatial rearrangement of primitive image objects, giving a smoother and more simplified view of the classified image. This image should, however, obey certain constraints, the most important being the local area accuracy, i.e. there must be a correct representation of the areas occupied by each class (using, for example a geostatistical kriging estimator based on local information). These area statistics are then used to drive the generalisation process.

A technique either used for pre-classification or post-classification processing is segmentation, which tries to detect the boundaries of relatively homogeneous parcels in order to remove small scale effects due to noise or classifier errors. At CVRM a set of segmentation algorithms were developed for grey level and colour images, based on the *watershed transform* enabling automatic detection of the homogeneous zones in the image from the stand point of spatial contiguity. Research on new segmentation methods to allow us to integrate the two main criteria present in the automatic classification of colour images (the spectral aspect or the representation domain of the attributes and the spatial aspect or the spatial contiguity domain of the same attributes) is ongoing.

During the last few years, other methodologies such as *knowledge based systems* (KBS) and *artificial neural networks* (ANN) have been widely used in the classification and post-classification of remote sensing data as a valid alternative to more conventional techniques. KBS can be considered as model-based systems, using simple geometric properties of spatial features and geographic context as rules, symbolic or evidential reasoning and expert systems architectures [1]. As contextual knowledge is of an uncertain nature and may only indicate tendencies, the KBS approach is a complex technique that requires skilled operators and specialised algorithms.

In contrast, ANN are a data-driven approach that have proved to be accurate and robust, even in the presence of noisy, fuzzy or incomplete data. ANN are universal approximators whose structure is inspired by the current understanding of biological nervous systems. They have the additional advantage of requiring a limited number of hypotheses regarding the nature and structure of the input data and they do not need any kind of *a priori* models, as they learn by example, they build their own internal representations of the system and are able to generalise in unknown situations. Consequently, their use in remote sensing automatic classification has increased considerably in the last few years and so has the list of references.

2. Artificial Neural Network Classifiers

ANN are computational models of a black-box type, composed of a large number of highly interconnected elements operating in parallel and arranged in one or more layers, where processing of information and memory are distributed amongst the whole structure. The connections have adaptive weights which are dynamically adjusted according to an optimising algorithm, and the processing elements (the artificial neurons or nodes) contain non-linear activation functions (typically a sigmoid). Furthermore, as a consequence of their highly redundant distributed structure, they exhibit robustness to incomplete, noisy or fuzzy data.

There are strong connections between ANN and statistical methods, as ANN can be regarded as non-linear regression analysis tools: their outputs are non-linear functions of a linear combination of the inputs; in consequence, they also present filtering properties. Their non-linearity allows them to solve problems that linear algorithms cannot solve. It has been demonstrated by comparison with statistical methods that a multilayer network showed a good potential for processing multi-source remote sensing and geographical data [5].

ANNs do present other drawbacks such as slow training time which is a function of the complexity of the network and of the data which can cause problems with convergence. They do however converge to an optimal or suboptimal solution. On the other hand, once they have been trained they are extremely fast in the generalisation phase.

Among the numerous models, the *multilayer feed-forward neural network* (MLFN) trained with the *back-propagation algorithm* (or generalised delta rule) [6] is by far the most commonly used paradigm for pattern recognition and classification problems, despite some drawbacks that they present related to the crucial training phase. The back-propagation algorithm is quite simple and can be implemented very easily. Its objective is to determine the set of weights that minimise the objective function, usually the mean square error between the actual and the desired outputs: it is a simple stochastic gradient descent algorithm, a class of the unconstrained non-linear optimisation methods [7].

The three main problems in the learning theory of MLFNs are: *the architecture problem* (how to choose the best architecture for the network), the *optimisation problem* (how to calculate the set of weights that minimise the error function) and the *convergence problem* (how to guarantee that a good performance can be reached in a reasonable time).

As is well known, in a multilayer network the size of the input and output layers are determined respectively by the dimensions of the input patterns (e.g. the number of bands) and the number of categories to be classified or the desired output mapping. The size of the hidden layer(s), however, is not specified *a priori* although this is a crucial point, as it allows the network to develop its own internal representation of the mapping and determines its memorisation/generalisation balance ability. Each specific application must therefore find the most appropriate architecture for the network to adequately solve the problem, although there is no theoretical basis for this choice.

The reduction of the training time is a crucial point in remote sensing image classification, because the enormous amount of data transforms the ANN training into a complex process, due either to the large number of the connection "weights" of the network that must be computed and adjusted in each iteration or to the large size of memory necessary. Improvements in training time can be achieved in two main ways with back-propagation: either using faster algorithms (or acceleration techniques for convergence), or reducing the dimensions of the problem. Another solution is the use of other paradigms. At the CVRM we made several attempts to deal with this problem and to improve performance.

There are some techniques to accelerate back-propagation that produce effective results and are described in the literature [8], such as those based upon the explicit or estimated evaluation of the diagonal components of the Hessian matrix, those based on the cosine of the angle of the gradient vector in two successive epochs, those that dynamically adapt the learning and momentum parameters, considering the signal changes of the gradient vector, and those originating in the conjugated gradient numerical techniques.

Recently, several neural network paradigms have emerged with emphasis on quick training; one of the most promising is the *Probabilistic Neural Network* (PNN), which was originally designed and formulated as an algorithm trained on members of two or more classes and used to examine unknowns to decide in which class they belong [9, 10]. The structure of ANNs and PNNs are topologically similar, but they are quite different in their operation modes.

Probabilistic Neural Networks are essentially classifiers based on the estimation of the joint probability density functions for each one of the pre-chosen classes (which must all have the same number of attributes). Several kernel functions can be used, but one of the most commonly used is the non-normalised Gaussian function, which has the advantage of having only one adjusting parameter (*sigma*). As a whole, PNN implements the Bayes decision rule, classifying according to the maximum estimated value of the probability density functions.

Like ANNs, PNNs are also linear combinations (simple summations) of the non-linear activation functions. They basically comprise three layers, a layer of input units, an intermediate or hidden layer of pattern units and a layer of output units (this third layer is usually subdivided into two, the last layer containing a single node which finds the largest output of the third). The input units correspond to the feature vector and distribute it to the pattern units, while the output units correspond to classes; pattern units hold the kernel functions and are equal in number to the training examples per class. Each pattern unit computes a distance measure between the input and the training case represented by that node and subjects that distance to the unit's activation function, which is basically a Parzen's window. The output units sum the activations of the pattern units corresponding to the respective unit's class and give the estimated value of the probability density function of that population; the decision of the network (performed in the fourth layer) is the class corresponding to the output unit with the maximum value.

Finally, the traditional method to simplify the training phase is the pre-processing of data, using any of the many techniques available. A promising approach under research at CVRM is a hybrid methodology conjugating statistical data analysis performed by *Principal Component Analysis* (PCA) with a stochastic search technique using *Genetic Algorithms*, which aims at reducing the amount of data in the initial training set, without important losses in performances [11]. The ANN classifier is trained with subsets of the training set, and the "best" ones are chosen through comparison of their classification rates with the performance reached with the whole training set.

Suppose there is a set with C classes to be classified according to k attributes or features. Firstly, the set is submitted to a PCA, and from each cluster formed in the plane of the first two principal axes, representing (theoretically) one class of patterns with k elements, the m ($m < k$) most representative samples are picked; therefore the ANN will be trained with mC samples instead of kC. However, this choice is the most critical part of the method: which m points will guarantee the most successful classification rate? The method used is to choose those furthest away from the others (the peripheral ones), because they can better represent the nuances of their classes. But, trying to prevent occasional overlapping, we select also a few points near the cluster barycentre, representing the most typical elements of each class.

For the determination of the peripheral elements we use a Genetic Algorithm, since a classical procedure is computationally expensive (and the objective of this approach is to reduce computing time). Genetic algorithms have been found to be robust and powerful optimisation methods in combinatorial search problems like this one. The problem is the inverse of the well known *travelling salesman problem*: the genetic algorithm punishes the genotypes of the population (representing the PCA distances between combinations of m images out of k) whose distances are smaller, and rewards the furthest ones, which go on the next generation by means of the standard genetic operators: selection, crossover and mutation.

The chosen subset (including a few barycentre points) is used to train the ANN. For comparison purposes some subsets formed by using the central points of each cluster (chosen by a centroid algorithm) and a subset of randomly picked points are also used as training data (all with the same number m of elements). The process finishes when the classification rate reached by the genetic algorithm subset is greater than the others and close enough to the performance of the whole set.

3. A Case Study

A typical and simple remote sensing study performed at the CVRM was the identification of the vegetation and soil cover in a rural area. The data set consisted of the digital radiances in 6 channels (blue, green, red, near-, mid- and far-infrared bands) of a Landsat-5 Thematic Mapper multispectral image of 1992, covering a rural region of Baixo Alentejo (South of Portugal) in a vast peneplain between Moura and Ficalho villages with a total area of $5460 \times 5250m$ (182×175 pixels) [12]; the image had a spatial resolution of $30 \times 30m$, with 8 bit values per pixel, i.e. 256 grey levels. These channels were chosen as the spectral reflectance of the vegetation, is higher in the infrared radiation zone.

The use of the standard digital classification techniques proved to be in-adequate due to the complexity of the vegetation cover characteristics, as commonly reported in the literature. So, the raw data was submitted to the usual corrections (radiometric, due to periodic stripping in the images, atmospheric, applying the minimum histogram method, and geometric, giving a geographical referential to the satellite images). Several types of soil covers (cornfields, ploughed and fallow grounds, bush, woody herbs, trees, etc) and water courses were identified in the area during field work.

Then a PCA was performed, using an interactive methodology with the maximum likelihood classifier (MLC) [12]. Finally, after grouping and removing some classes, those discriminated by PCA were classified by the MLC. In general, mathematical morphology operators are also used to filter the resulting classified images and suppress distortions induced by digital classification.

With the aim of testing the performance of the neural network approach compared to the traditional technique, we applied a MLFN and a PNN classifier to the same training set used by the MLC, with 5 soil pattern classes, and compared results of the three methods with those coming from field work, by means of confusion matrices [13, 14, 15].

The ANN was simulated by a computer program ("NETFOR") developed at the CVRM with modular structure, implementing the back-propagation model and using a sigmoid as transfer function. It was written in FORTRAN and runs under DOS on any Personal Computer of 486 type or superior. We used 6 elements in the input layer, corresponding to the radiances in the 6 channels, 5 in the hidden layer (chosen after some tuning tests) and 5 in the output, determined by the number of training classes, and adaptive learning parameters.

The PNN program was simulated by a program in the C++ language implementing a four layer network (the 3rd layer was subdivided into two with the 4th layer automatically giving the class with higher membership probability) and using a Gaussian function as a kernel. It could use either a model with separate sigma parameters for each variable, but shared among all classes, or different sigmas for each variable and class.

4. Results

The overall classification accuracies given by confusion matrices in the test areas were, 83.6% for the MLFN, 91.9% for the PNN, and 83.0% for the MLC. Generally speaking, the discrimination between the 5 classes was satisfactory and classification of water courses, cornfields and sowing soils seems similar in the three methods, but distinction between ploughed and fallow grounds was nearly always difficult. Both the MLFN and the PNN approaches to soil classification give good performance: the best results in all cases are those of high density woody herbs and trees; the worst results correspond to soils masked with creeping vegetation, some trees, etc. and to wet lands and flooded areas or border zones between different classes, giving in some cases "no classified" pixels (those corresponding to a membership probability less than 0.5) and showing the need for prior image enhancement.

The performance of the PNN classifier was the best, both in terms of accuracy and computing time. We are currently developing other applications of this paradigm with more complex and larger remote sensing classification problems, in order to confirm (or invalidate) the capabilities shown in these simple tests.

References

1. G. G. Wilkinson, "Reasoning with Geographical Knowledge in Satellite Image Understanding", in: *Proceedings EUROSTAT/DOSES Programme Workshop on New Tools for Spatial Data Analysis, Lisbon 17–20 November 1994*, Eurostat Statistical Document, Theme 3, Series D, Office for Official Publications of the European Communities, Luxembourg, pp. 109–119, 1994.
2. F. Muge and P. Pina, "Application of Morphological Operators to Supervised Multidimensional Data Classification", in: *Mathematical Morphology and its Application to Image Processing*, J. Serra and P. Soille eds., pp. 361–368, Kluwer Academic Publ., 1994.
3. F. Muge and P. Pina, "Some Applications of Mathematical Morphology Based Image Analysis Techniques to Remote Sensing", *Proceedings of RECPAD'93 Conference*, pp. 287–294, Porto, 1993.
4. J. Serra, *Image Analysis and Mathematical Morphology*, vol. 2, Academic Press, 1988.
5. J. Benediktsson, P. Swain and O. Ersoy, "Neural Network Approaches versus Statistical Methods in Classification of Multi-source Remote Sensing Data", *IEEE Transactions on Geoscience and Remote Sensing*, vol. 28, no. 4, pp. 540–552, 1990.
6. D. Rumelhart, G. Hinton, and R. Williams, "Learning Internal Representations by Error Back-propagation", in: *Parallel Distributed Processing. Explorations in the Microstructure of Cognition*, D. Rumelhart, J. McClelland and the PDP Research Group eds, vol. I (Foundations), pp. 319–364, MIT Press, 1986.
7. A. Cichocki and R. Unbehauen, *Neural Networks for Optimization and Signal Processing*, Wiley, 1993.

8. F. Silva and L. Almeida, "Speeding up Back-propagation", *Advanced Neural Computers*, pp. 151–158, North Holland, 1990.
9. D. Specht, "Probabilistic Neural Networks and the Polynomial Adaline as Complementary Techniques for Classification", *IEEE Transactions on Neural Networks*, vol. 1, no. 1, pp. 111–121, 1990.
10. T. Masters, *Advanced Algorithms for Neural Networks, A C++ Sourcebook*, Academic Press, 1995.
11. V. Ramos, "Using Principal Component Analysis and Genetic Algorithms Techniques to Optimize Learning Time on Neural Network Training for Pattern Recognition", G.I.A.I. Notes no. 1/VCR/96, CVRM ed., Lisbon, 1996.
12. T. Barata "Vegetation Cover Identification in Landsat TM Images by Digital Classification and P.C.A.", *Proceedings of RECPAD'94 Conference*, pp. 193–200, Lisbon, 1994.
13. L. Cortez, F. Durão, T. Barata, C. Guimarães and P. Pina, "Neural Network Techniques Applied to Classification of Digital Images", *Proceedings of RECPAD'94 Conference*, pp. 29–36, Lisbon, 1994.
14. L. Cortez, F. Durão, and T. Barata, "A Neural Network Approach to Discriminate Thematic Images of Rural Areas", *Proceedings of RECPAD'95 Conference*, 7.4.1/7.4.12, Aveiro, 1995.
15. L. Cortez, F. Durão, and T. Barata, "Application of Probabilistic Neural Networks to Classification of Digital Images", *Proceedings of RECPAD'96 Conference*, pp. 407–414, Guimarães, 1996.

Using Artificial Recurrent Neural Nets to Identify Spectral and Spatial Patterns for Satellite Imagery Classification of Urban Areas

Sara Silva and Mario Caetano *

CNIG (*Centro Nacional de Informação Geográfica,*
National Centre for Geographic Information,)
Rua Braamcamp, 82, 1° Dto, 1250 Lisboa, Portugal.
e-mail: {sara,mario}@cnig.pt

Summary. The majority of techniques used for satellite imagery classification usually perform poorly on discriminating urban land use classes, either because they have similar spectral signatures or because the patterns they exhibit are broader than satellite image pixels. In this paper we tackle a new classification methodology, based on spectral and spatial pattern analysis using artificial neural networks. First a self-organising classifier splits the spectrum of individual pixels on spectrally pure land cover classes. Then a second classifier self-organises critical regions of adjustable topology on the resultant image, and automatically classifies it into land use classes. Both classifiers are implemented on artificial recurrent neural networks inspired by Carpenter and Grossberg's adaptive resonance theory (ART).

1. Introduction

The majority of techniques used for satellite imagery classification usually perform poorly on discriminating urban land use classes, either because they have similar spectral signatures or because the patterns they exhibit are broader than satellite image pixels.

For the first case we may mention the confusion between spectral signatures of residential areas and industrial or commercial areas. One single pixel does not contain enough data to allow its classification, because the discrimination of these classes can only rely on factors such as the homogeneity of the surrounding area. In the second case we refer to the problem of discriminating different types of residential area, which strongly depends on the identification of patterns formed by the superposition of green and urban areas.

One of the main reasons for these two problems can be the fact that the satellite imagery classification is usually performed at a pixel level and consequently classes whose discrimination depends on spatial analysis are not identified. Also, because of the great variety of mixtures found, the performance of the more traditional supervised methods is handicapped by the great difficulty of finding suitable training areas. Our classification methodology is based on spectral and spatial pattern analysis using artificial neural

* This work was supported by a grant from CNIG and by the project COMBINA (JNICT PBIC/TIT/2527/95)

networks. First a self-organising classifier splits the spectrum of individual pixels on spectrally pure land cover classes. Then a second classifier self- organises critical regions of adjustable topology on the resultant image, and automatically classifies it into land use classes. Both classifiers are implemented on artificial recurrent neural networks inspired by Carpenter and Grossberg's adaptive resonance theory (ART) [1, 2, 3]. We use ART1 for the first classification task and ART2 for the second.

The methodology is tested with a SPOT image of *Área da Grande Lisboa*, Portugal. Besides the three digital numbers (DNs) between 0 and 255, belonging to the SPOT bands XS1, XS2 and XS3, we also use the *normalised difference vegetation index* (NDVI).

As this work was our very first attempt at using ART networks for satellite imagery classification, some of the results presented here are more qualitative than quantitative. In our first approach, our major goal was to study the ART network's behaviour on satellite imagery classification tasks, and not to pursue the best possible results.

2. Adaptive Resonance Theory

ART architectures constitute an hierarchy of biologically inspired neural networks used to self-organise arbitrary sequences of input patterns. Their major feature is the ability to overcome Grossberg's *stability-plasticity dilemma*, according to which a neural network should be able to learn new input patterns (*plasticity*) without affecting the storage and recalling of the previously learned ones (*stability*). ART achieves this by relying on a vigilance parameter which determines whether the new input pattern should be accepted as a member of an existing category (class) or, on the contrary, adopted as the prototype of a new category. If this is the case, but the network storage capacity does not allow the creation of new categories, the new pattern is simply rejected, thus preserving the information the network has already acquired.

ART network operation follows naturally from their architectures. ART1 is designed to self-organise only binary input patterns, whereas ART2 is able to deal with either binary or analogue ones. Their dynamics can be fully described by differential equations.

ART1

Figure 2.1 illustrates the main components of an ART1 network. Rectangles represent layers and circles represent nodes. Thin lines represent specific links and thick lines represent non-specific excitatory (arrow) or inhibitory (circle) links, or connections. Layers I and F1 are the external and internal input layers, respectively, and F2 is also called the output or storage layer.

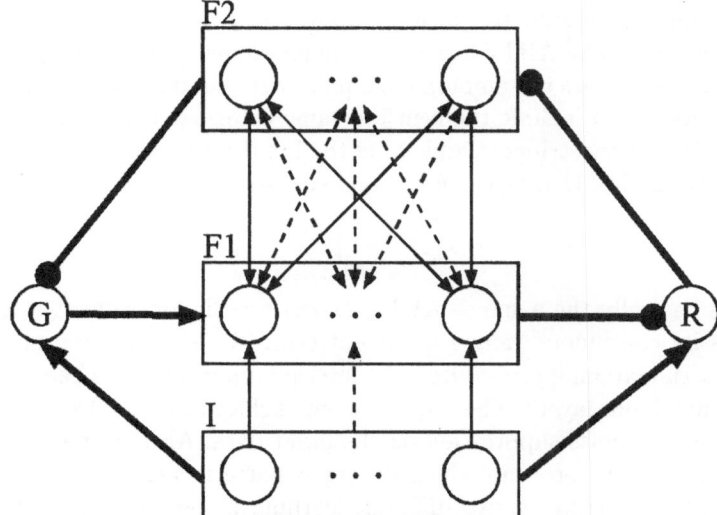

Fig. 2.1. Architecture of the ART1 network.

The specific links form a one-by-one connection between the nodes of layers I and F1 and two separate full connections (*feed-forward and feedback*) between the nodes of layers F1 and F2. The goal of the network is to classify binary input patterns into categories, or clusters, created and associated at run-time to the nodes of the storage layer.

The one-by-one connection is responsible for passing each external binary input pattern received by layer I to layer F1. Three possible signals are likely to arrive at layer F1 - the excitation from node G and the ones arriving from layers I and F2 - but only when activated by two of them, layer F1 sends its pattern through the forward links to F2. This is denominated as the *2/3 rule*.

Although not illustrated in the figure, the output layer forms a competitive Winner-Take-All (WTA) structure, where each node excites itself and inhibits all its neighbours. As a result, only one of the output nodes is active at a time, representing the category to which the current input pattern belongs. When node R inhibits layer F2, the current winner node is disabled and removed from the competition until a new input pattern is to be categorised. This particular inhibition is called a *reset signal*. The feedback links between layer F1 and each output node store the representation of the category associated to the node. The forward links represent its normalisation.

The initial feedback weights must be such that the competition in the output layer is not dominated by a single node and any first time winner is not disabled by a reset signal. The forward weights must be initialised with lower values than the ones reached after resonance. Carpenter and Grossberg chose to initialise the feedback weights with the value 1 and the forward weights with $\frac{1}{(1+n)}$, where n is the number of input nodes.

The vigilance parameter, ρ, must obey $0 < \rho \leq 1$ and can be changed during learning. The ART1 learning is highly dependent on this parameter, because it determines the amount of similarity required between two patterns for the network to include them in the same category. If ρ is high, the network becomes very stringent and splits the input patterns into various finely divided categories. Otherwise, fewer and coarser categories are created.

ART2

ART2 is basically the same as ART1, except for the fact that it can accept and classify continuous valued input patterns. To deal with the more complex pattern matching procedures, the internal input layer has been split into several functional layers. The improvements achieved by ART2 include normalisation and noise suppression on the input data. Also, its performance is not affected by the order in which the input patterns are presented.

ART2 can operate in two different learning modes. Fast learning needs only one iteration to optimise the weight values. It is appropriate for binary input patterns only. Slow learning, on the contrary, is advisable for analogue inputs. In this mode, the weight values are adapted over many iterations, thus achieving the representation of each category's statistical average.

3. Data Representation

Given the network description, one can easily understand that ART1 can be used for spectral classification of urban areas, as long as we can find an appropriate binary codification for the satellite data. The one we have used is very straight. It simply splits the range of the 256 possible satellite DNs into a predefined number of intervals, not necessarily of equal sizes and not necessarily the same for all the satellite bands. Each DN then falls into only one interval, which is set to 1, while the others remain 0. Grouping the representations of the three SPOT DNs available for each pixel, we obtain the binary patterns that will feed the network. Figure 3.1 illustrates the binary codification of one pixel whose DNs in bands XS1, XS2 and XS3 are respectively 50, 70 and 150, with each band split into four equal sized intervals.

The NDVI is obtained from the DNs of bands XS3 and XS2, dividing its difference by its sum $\frac{(dnXS3-dnXS2)}{(dnXS3+dnXS2)}$. The result is a continuous valued number between -1 and 1, for each pixel, which we have converted to discrete numbers between 0 and 255. The NDVI binary codification is the same as the DNs, but here we have used non equal fixed length intervals which have already been proven to achieve good discrimination of urban land cover. Those are 0–109, 110–130, 131–150, 151–175 and 176–255, therefore five intervals.

As already mentioned, ART2 is used for the spatial analysis of a previously spectrally classified image. Therefore, each pixel's input pattern contains data

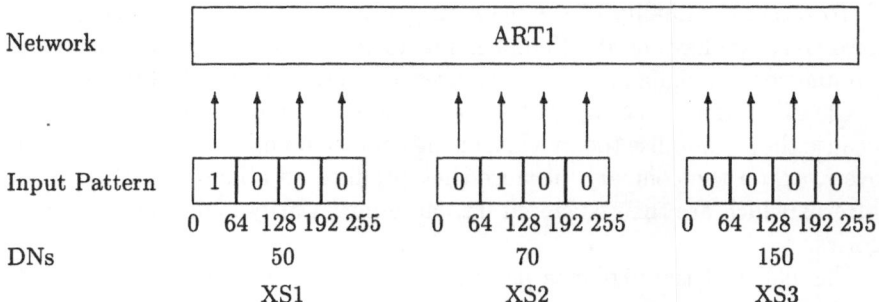

Fig. 3.1. Binary representation of one pixel with DNs 50, 70 and 150 in bands XS1, XS2 and XS3, using four equal sized intervals for each band.

concerning a whole window, or kernel. Needless to say, when the kernel size is reduced to a single unit, the classification performed is purely spectral. ART2 input patterns are vectors containing the percentages of each class relative to the total number of classified pixels inside the kernel. Unclassified pixels are tolerated in this way. Figure 3.2 illustrates the representation of the middle pixel of a 3 × 3 kernel containing one unclassified pixel and equal percentages of vegetation and urban.

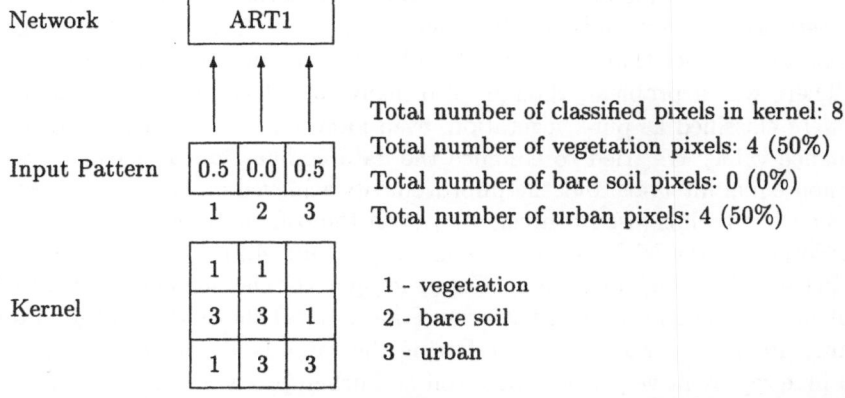

Fig. 3.2. Representation of the middle pixel of a 3 × 3 kernel with one unclassified pixel and equal percentages of vegetation and urban.

4. Results and Discussion

In spite of all the spectral variability usually present, three general types of land cover constitute almost all that can be found in urban areas: vegetation, bare soil and urban. Knowing that, we have selected a few small areas (475 pixels total) of those known cover types and fed them to ART1. The input data was solely composed of the three SPOT bands' DNs.

To assess the quality of the network performance on the classification of those areas, we have used different numbers of intervals in the data codification and several vigilance values. The results were as shown in table 4.1.

Clearly, some of the input data representations are unacceptable, either because they generalise too much, causing confusion between the various land covers, or, on the contrary, because they produce an abundance of different patterns which are then clustered into an unnecessarily high number of categories.

The effect of the vigilance parameter on the ART1 performance is also clearly demonstrated. Low vigilance values tend to cause the inclusion of different land covers into the same cluster, because the network does not find them different enough to treat them as separate categories. High values, on the contrary, produce too many similar clusters.

From all these classifications, the one using five intervals and a vigilance value of 0.6 was the one yielding better results. Nevertheless, both bare soil and urban were split into two different clusters each, revealing their enormous spectral variability.

Using this particular result, we tested ART1's newly acquired knowledge on a large (3392 pixels) area including pixels from all typical urban classes. The network classified 98% of the image. The remaining pixels, spread all over the area, were considered unknown. Curious about their nature, we allowed the network to learn and classify them. They were clustered into so many different categories that we preferred to let them remain unclassified.

There was a problem, though. Too many mixed vegetation/urban pixels were classified as pure vegetation, even after a dramatic increase of the vigilance value. We tried to enhance the data codification by means of using non equal intervals, but the improvements achieved on these pixels were crushed by the increased confusion between the others. So we have decided to incorporate the NDVI into the input data, (table 4.2).

In spite of the high number of clusters formed, the classified image revealed a lot more vegetation/urban patterns. Figure 4.1(a) shows this image, after joining all the separate clusters referring the same land covers. Red, green and blue represent vegetation, bare soil and urban pixels, respectively. Black means unclassified.

Still, most mixed pixels were classified as pure vegetation. Because the NDVI had made a lot of difference, we decided to use it, alone, to perform another classification of the same image. This time we used ART2, in fast learning mode and with a vigilance value of 0.7, with a unitary kernel size - a pure spectral classification. The results are shown in the image of figure 4.1(b) and were truly surprising: all mixed pixels were classified as urban, entirely revealing the typical vegetation/urban patterns. No attempt was made to reduce the number or interpret the classes produced by the NDVI classification.

Table 4.1. Classification results using different interval and vigilance values.

# Intervals	Vigilance	# Clusters	Cluster	# Pixels	Land Cover
4	0.6	5	1	139	1
			2	2	1
			3	136	2
			4	5	2
			4	192	3
			5	1	3
4	0.9	8	1	136	1
			2	3	1
			3	2	1
			4	124	2
			5	12	2
			6	4	2
			7	1	2
			6	192	3
			8	1	3
5	0.3	3	1	141	1
			1	64	2
			2	77	2
			3	193	3
5	0.6	5	1	141	1
			2	77	2
			3	64	2
			4	188	3
			5	5	3
5	0.9	10	1	139	1
			2	2	1
			3	20	2
			4	57	2
			5	16	2
			6	48	2
			7	181	3
			8	7	3
			9	3	3
			10	2	3
6	0.6	9	1	139	1
			2	2	1
			3	23	2
			4	109	2
			5	4	2
			6	5	2
			1	18	3
			6	60	3
			7	77	3
			8	36	3
			9	2	3

Table 4.2. Classification results using NDVI in the input data.

# Intervals	Vigilance	# Clusters	Cluster	# Pixels	Land Cover
5	0.6	9	1	139	1
			2	2	1
			3	76	2
			4	63	2
			5	1	2
			6	1	2
			7	181	3
			8	9	3
			9	3	3

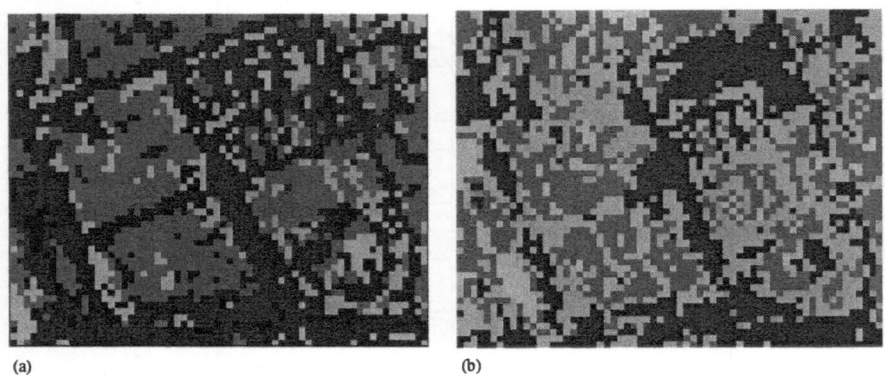

(a)   (b)

Fig. 4.1. (a) Spectrally classified image by ART1, using five intervals, a vigilance value of 0.6 and the incorporation of the NDVI in the input data. Red, green, blue and black represent vegetation, bare soil, urban and unclassified pixels, respectively. (b) Spectrally classified image by ART2 in fast learning mode with a vigilance value of 0.7, using the NDVI alone.

The ART2 spatial classification capabilities were then tested on both images, with kernel sizes ranging from 3 × 3 to 9 × 9. Smaller kernels produced better images. Figure 4.2 illustrates the advantages of the ART2 slow learning mode. Both images were classified using a kernel size of 3 × 3, but the most accurate is the left one, classified in slow learning mode. The difference becomes even bigger if we consider that the classification of this image was performed on a worse input data set, i.e., the classified image in figure 4.2(a).

5. Conclusion

We have used ART networks on satellite imagery classification and studied their behaviour on the classification of urban areas, where a high quantity and mixture of classes is found. As our very first approach, we have mainly focused on the use of spectrally classified images by ART1 as the input data

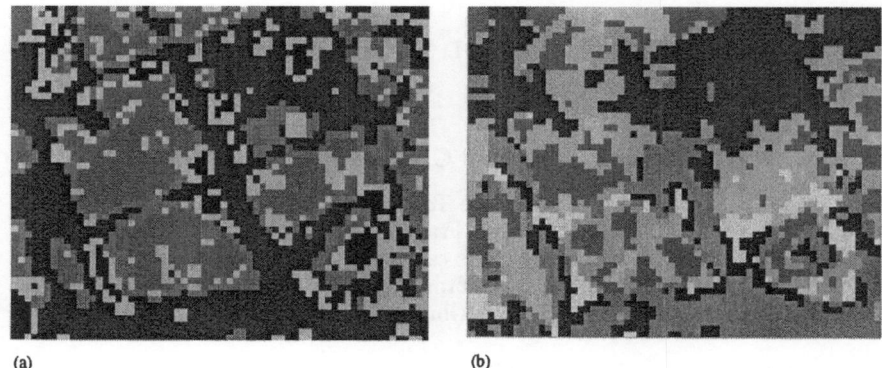

(a) (b)

Fig. 4.2. (a) Spatial classification of the image in figure 4.1(a) by ART2 with slow learning. (b) Spatial classification of the image in figure 4.1(b) by ART2 with fast learning.

for the spatial classification of ART2. Because the spatial classification performance is thus entirely dependent on the accuracy of the previous spectral analysis, in the future we intend to use the more powerful ART2 on both classifications. The combination of input data from various sources, along with its pre-processing and codification is an open field for future investigation.

Acknowledgement. We would like to thank Dr. Josè Felix Gomes da Costa for sharing so much of his neural expertise with us.

References

1. G. A. Carpenter, "A massively parallel architecture for a self-organising neural pattern recognition machine", *Computer Vision, graphics and image processing*, vol. 37, pp. 54–115, 1987.
2. S. Grossberg, "Studies of Mind and Brain: Neural Principles of Learning", *Perception, Development, Cognition and Motor Control. Boston Studies in the Philosophy of Science*, D. Reidel Publishing Company, vol. 70.
3. S. Grossberg, "Competitive learning: From interactive activation to adaptive resonance", *Cognitive Science*, vol. 11, pp. 23–63, 1987.

Dynamic Segmentation of Satellite Images Using Pulsed Coupled Neural Networks

X. Clastres[1], M. Samuelides[1,2], and G. L. Tarr[3]

[1] Centre d'Etudes et de Recherches de Toulouse
ONERA, BP 4025, 31055 Toulouse, France.
e-mail: clastres@cert.fr, samuelid@tls-cs.cert.fr
[2] ENSAE, BP 4032, 31055 Toulouse, France.
[3] Phillips Laboratory, US Air Force, Albuquerque, New Mexico, USA.

Summary. Pulsed Coupled Oscillatory Neural Networks are examined for application to image analysis. Adapting to biological constraints, a pulsed coupled network using an Integrate and Fire model with dynamical synapses is examined to perform image segmentation based on synchronisation on the firing time of neurons which are in the same region. To enhance synchronisation behaviour an avalanche-type dynamic is introduced. This dynamic allows us to perform segmentation by synchronisation in the transient regime, without having to reach a stable stationary regime which is expensive in terms of computation time. Using the pulsed time-of-arrival as the information carrier, the image is reduced to a time signal which allows an intelligent filtering using feedback. A multi-layer implementation of the model is presented that shows good segmentation results and is easily adaptable to multispectral imagery.

1. Introduction

The current view of the world from space is like looking through a soda straw. Without automated processing, much of the information collected by space resources as the SPOT and Landsat satellites are lost for lack of storage and people to evaluate them. As yet, most automated image analysis techniques are not robust enough to be applied across a wide variety of imagery collected from space. Conventional image processing techniques which work for one class of imagery may not work for another class of imagery. Effective and efficient methods to process massive amount of image data might be discovered if we examine biological systems. Actually, the problem represented by analysis of satellite multi-spectral imagery is similar to that faced by biological systems: massive throughput of data, multiple spectral channels and complex classification of objects. An interesting difference between biological vision and computer vision systems is the extensive use of feedback. In this paper we will examine the use of feedback techniques for low level image processing methods. The biological relevant paradigms are recalled in section 2, our model is describe in section 3. Then various results of simulations are presented showing the functionalities of the system. We end by a short discussion on further improvement and application of our model.

2. Pulsed Based Neural Networks

Gray (1987) and Eckhorn (1988) [1, 3] have observed, in the visual cortex of the cat, stimulus related oscillations of neuron activity. They suggest that information transport in the biological vision system may be based on the observed synchronisation between pulses of different neuronal clusters. Pulsed oscillatory systems are presented here as a paradigm for image processing.

The main difference between dynamic spiking neurons and analogic neurons (commonly used in image processing) is that the information is not only coded with the firing rate of the neuron but also in terms of phase. Two neurons that have the same frequency of discharge can be differentiated by their different phase. The most know of them is the Integrate and Fire model (IF).

2.1 The Eckhorn PCNN Model

In order to explain the mechanism of synchronisation, Eckhorn has proposed [3] a model which reproduces some of this behaviour. It is an integrating model with dynamical synapses.

The neuron consists of a membrane potential U and a threshold θ. When, at time t, the neuron potential $U_j(t)$ reaches the threshold value $\theta_j(t)$, an action potential is emitted by the neuron. The output of neuron j is the binary variable $Y_j(t)$ which takes the value $+1$ when the neuron fires and the value 0 otherwise. The threshold behaves like a leaky integrator, it exponential decreases to an offset value θ_0 with a time constant τ_θ. When a spike is produced there is an increase in threshold by a fixed value θ_s. Two regimes can be observed depending of the liking strength and the θ_s value. The first allows synchronous bursts of pulses. In the second, neurons are prevented from firing in burst by adding a refractory period to the neuron. The refractory period is modelled by a high enough θ_s value preventing the neuron from firing for a minimum time. The neuron is also characterised by its dendritic tree, which has two distinct branches: the Feeding and the Linking. The Feeding $F(t)$ is the main input from the captor input (stimulus) and from neurons of lower layers. It represents the direct signal longitudinally propagated. The Linking L(t) is the modulating signal providing global information from the neighbourhood. Each of the two dendritic branches is dynamical: leaky integrator with respective time constants α_F and α_L. The model is described by the following equations:

$$\begin{cases} U_j = F_j \cdot (1 + \beta \cdot L_j) \\ L_j(t+1) = \sum_k (W_{kj}^L \cdot e^{\frac{t}{\alpha_L}}) \star Y_k(t+1) \\ F_j(t+1) = \sum_k (W_{kj}^F \cdot e^{-\frac{t}{\alpha_F}}) \star Y_k(t+1) + X_j(t+1) \end{cases} \qquad (2.1)$$

where W_{kj}^L and W_{kj}^F are the synaptic coefficients of the two different branches, connecting neurons j and k. $\star$ is the convolution product. β is the linking strength.

$$Y_j(t+1) = \begin{cases} 1 & \text{if } U_j(t+1) \geq \theta_j(t+1) \\ 0 & \text{otherwise} \end{cases} \tag{2.2}$$

$$\theta_j(t+1) = \theta_0 + (\theta_s \cdot e^{-\frac{t}{\alpha_\theta}}) \star Y_j(t+1) \tag{2.3}$$

Built into this neuronal model are a number of characteristics common to biological vision: integrate and fire, latency and lateral interconnectivity. The following graphs provide a means to visualise the architecture and internal behaviour of the Eckhorn neuron.

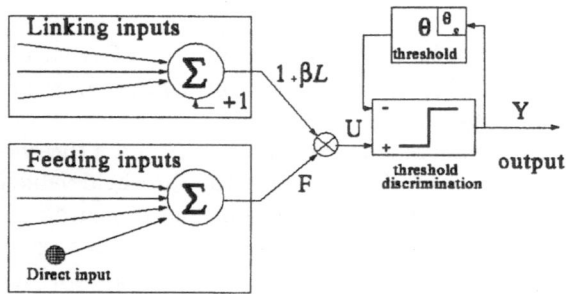

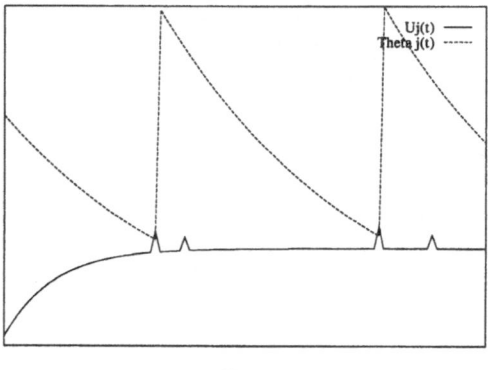

Temps

The first graph represents the schema of the tree architecture of the neuron. Each synapse of the Feeding and the Linking inputs is a leaky integrator. In the second graph, the threshold and membrane potential diagram represents the behaviour of these variables and shows the pulse capture by linking signal. The energy is integrated and plotted. The threshold, a decaying exponential, relaxes from its initial value. When the neuron potential reaches the threshold, the neuron fires and a fixed amount θ_s is added to θ. This ensures a minimum pulse rate and a refractory period (for a minimum value of θ_s).

3. Transient PCNN for Image Processing

Pulsed Coupled Neural Networks (PCNN) like Eckhorn neurons have been introduced to explain the experimentally observed synchronous activity among neural assemblies in the cat cortex. This model has shown good capacity to perform image processing. Actually, neurons that are associated to regions of common textures or characteristics in the visual field will share the same common pulse rate because of the feeding, and be synchronous in time under the effect of the linking field. Given these characteristics, several image processing functions are possible ([4, 5]). In this paper we will explore one of these functions, generally accepted as the most difficult: segmentation. In [4] Johnson used this model to achieve segmentation by synchronisation of the neuron pulses; the synchronisation measure is estimated from the stationary periodic regime. But the synchronisation of the neurons is not complete; when some close-by neurons belonging to different regions are firing almost simultaneously, the linking effect may induce shifts in the period of some neurons. Another drawback of this stationary regime analysis is the cancellation of the information from the transient regime and that means additional computation time to reach a stable stationary regime. Furthermore, Thorpe [6] notes that it is very unlikely that enough time exists to reach such a stationary state for such synchrony given how quickly a real image is processed by the brain. He suggest that not only synchrony, but also time of arrival are important in determining regional features. So, we use the latency delay to replace the rate of discharge to achieve image segmentation. This was made possible by implementing a new synchronous dynamic on our network.

3.1 The Dynamics

The dynamics of our network differs from the classical synchronous dynamic used in the Eckhorn model. The coupling of equations (2.1) and (2.2) have to be managed according to a computational dynamic. The "avalanche synchronous" dynamic gives the best results for our synchronisation purpose. This dynamic have been introduced by Bak et al. in [7] and is extensively studied in theoretical physics for its nice self-organisation properties. It is characterised by a double scale: the time scale of iterations (i.e. time t) and a communication scale between neurons. We assume that communication time between neighbour cells is short compared to the time step of iterations. Namely, when a neuron is connected to "ripe" neurons, the firing of this neuron will instantaneously induce the firing of the connected ripe neurons, and this process of avalanche will be repeated without changing either the threshold or the feeding inputs as many times as needed. All neurons firing in the same avalanche are, due to the relation between time scales, synchronous. This dynamic gives excellent results in smoothing the noise without blurring the edges.

Another interesting feature is that the segmentation is translated by a data flow allowing further real time higher level processing: feedback control of a camera, mathematical morphology operators and so on.

3.2 The Satellite Image Processing

This section describes how Eckhorn's neuron can be modified and used for image processing as smoothing noise, histogram equalisation or segmentation. To perform these tasks, PCNN we use differs from Eckhorn network in the following points:

- In the network the number of neurons is equal to the number of pixels of the input image. Each of them is centred on a pixel of the image.
- The direct input of a neuron is a translation of grey level of the associated pixel. Feeding input provides a fusion of information from several features in the recepting field (local region around the direct input).
- For image segmentation applications we use a linearly decaying threshold. The decaying step is noted Δ_θ. This evolution rule gives uniform processing on high and low amplitude stimulus regions.
- Each neuron receive a Linking input from neurons in its neighbourhood. This neighbourhood forms a R radius disc centred on the neuron. The linking strength of the connection between neuron i and neuron j is inversely proportional to the distance between them.

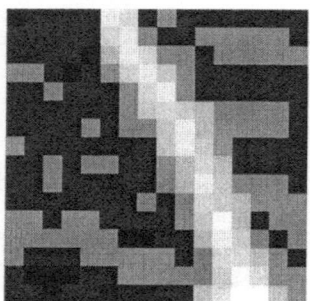 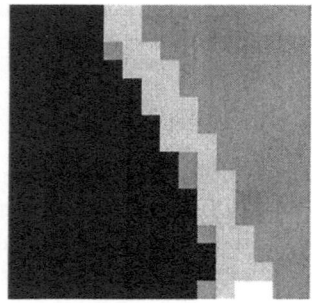

The above 16x16 image subset is taken from a SPOT satellite scene and represents a common problem for image segmentation. Most common edge detectors, neighbourhood averaging or median filtering methods use convolution filter techniques. The resolution of these methods are reduced to the size of the filter window. The neighbourhood averaging method blurs edges.

Median filtering methods erode or dilate edges. Simulation studies shows that PCNN is able to segment or smooth noise without blurring, eroding or dilating edges. An important advantage of our method is that segmentation is possible for objects smaller than the window size which is here the range of the linking effect. In the considered image subset a road is detected having a width in some places of less than two pixels. General threshold setting for this problem is not compatible with detection of relevant features.

The left image above is taken from SPOT scene of French countryside which is divided by various factors: property, land, culture, roads, village and so on. Segmentation of these portions is difficult. The edges have to be respected in order to perform further sequences of image fusion. Using radiometric information, the PCNN processing binds neurons corresponding to homogeneous regions of the image. However, the pulse capture mechanism just allows a spike to be anticipated. The more excited neurons will fire first then may induce the anticipated firing from its neighbours. Then, this processing will poorly perform on a dark zone of the image with bright dots in it. To avoid this drawback, inhibitory linking has been introduced slowing down the activity of isolated spiking neurons. The following images show the effect of this mechanism.

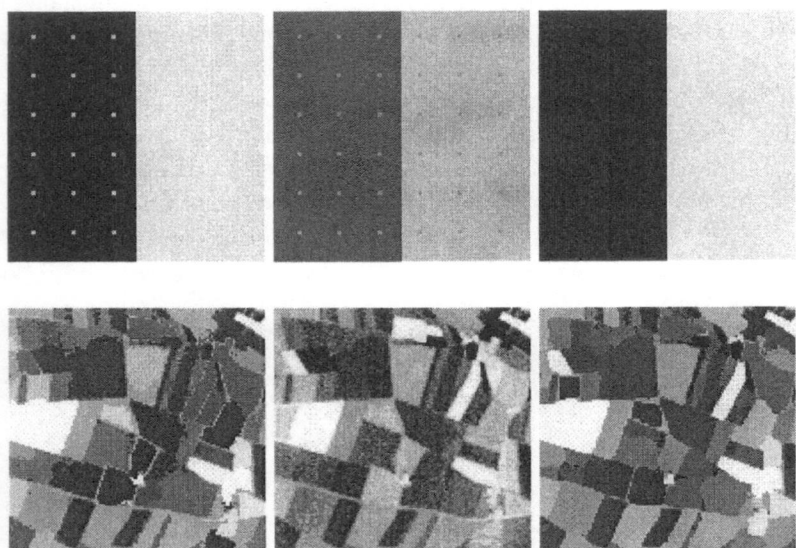

Left: processing of excitatory cells. Centre: input images. Right: Processing using both type of connections.

Most of the common techniques perform relatively poorly on this type of imagery. The advantage with this new technique is that there are almost no parameters to be set, and no hidden decision made by the operator. The algorithm was applied to the image directly without much adjusting of tuning. Such a simple architecture, is of course not able to provide an optimal result at the first shot. In particular, there is no symmetry between the processing of dark zone and clear zone. That suggests to fuse the two separated processing stages for the image and the inverse image which is also biologically plausible. In addition, through the feeding input, a neuron can extract given features from the input image or work on different types of information. Then parallel layers can process different treatments on the same image. These parallel cooperating processings can then cooperate and be integrated to perform better segmentation.

4. Conclusion

This paper documents the first efforts of a joint research between the United States Air Force Phillips Laboratory and the Office National d'Etude et de Recherches Aérospaciales. The models developed show effective image segmentation over a wide variety of imaging problems. While many of the results demonstrated here could be duplicated using standard techniques, these new methods offer a simple modular approach to the image analysis. An important result of this work is the statement that dynamic mechanisms are possible in biological vision systems to perform a type of histogram analysis of the imagery.

The next step of this work will be to introduce several layers of detectors and to fuse their information in order to perform segmentation on Multi-Spectral Imagery. Another important issue of this research is the translation of these techniques to silicon devices. Phillips Laboratory in cooperation with ONERA/CERT is working to design integrated circuits with automatic histogram equalisation and segmentation built in.

References

[1] C. M. Gray and W. Singer, "Simulus-dependent neuronal oscillations in the cat visual cortex area", *2nd IRBO Congress,Neuroscience Suppl.*, 1301P, 1987.

[2] R. Eckhorn, R. Bauer, W. Jordan, M. Brosch and W. Kruse, "Coherent oscillations: mechanism of feature linking in the visual cortex? multiple electrode and correlation analysis in the cat", *Biological Cybernetics*, vol. 60, pp. 121–130, 1988.

[3] R. Eckhorn, H. J. Reitboeck, M. Arndt and P. Dicke, "Feature linking via synchronisation among distributed assemblies: Simulations of results from cat visual cortex", *Neural Computation*, vol. 2, pp. 293–307, 1990.

[4] J. L. Johnson and D. Ritte, "Observation of periodic waves in a pulse coupled neural network", *Optics letters*, vol. 18, no. 15, pp. 1253–1255, 1993.

[5] J. L. Johnson, "Time Signature of Images", *IEEE International Conference on Neural Network,Orlando,FL*, 1994.

[6] S. Thorpe and M. Imbert, "Biological constraints on connectionist modelling", in: *Connectionism in Perspective*, Elsevier Science Publishers, North Holland, 1989.

[7] P. Bak, C. Tang and K. WiesenfeldK, "Self organised criticality: An explanation of 1/f noise", *Physical Review Letters*, vol. 59, no. 4, pp. 381–384, 1987.

Non-Linear Diffusion as a Neuron-Like Paradigm for Low-Level Vision

M. Proesmans, L. J. Van Gool, and P. Vanroose

ESAT-MI2, Katholieke Universiteit Leuven,
Kard. Mercierlaan 94, B-3001 Leuven, Belgium.

Summary. Remote sensing puts high demands on image processing. It calls for state-of-the-art algorithms, e.g. neural networks. However, neural nets usually work on preprocessed data and the preprocessing steps themselves have proved difficult to implement with NNs. Here a NN-like paradigm for low-level image processing is presented, that is based on the evolution of coupled, non-linear diffusion equations. The illustrations are focussed on feature preserving noise reduction, but the framework is more general.

1. Introduction and Motivation

By their very nature, satellite and aerial images are difficult to process robustly on the basis of computer vision. The scenes are complex and there are many sources of irrelevant variability. Reducing such noise without destroying meaningful features has been a longstanding research issue. Crop classification rates might e.g. go up when smoothing the available multispectral information over several pixels in the same field, but at the same time performance drops as data get smeared out over field boundaries. This should be avoided. Simple linear convolution filters are ruled out from the start then, and non-linear filtering is called for.

We propose to use systems of coupled, non-linear diffusion equations to solve a host of low-level computer vision problems. Such systems share the following characteristics with biological neural processes:
1) they are amenable to massive, fine-grained parallelism,
2) the result emerges from interconnected, local processes,
3) processing is split over specialised modules (areas in the brain, here equations),
4) these modules are typically bidirectionally coupled.
These points are clarified further on. An important difference with several *artificial* neural networks is the *local* connectivity that suffices between the "cells", here the computational units at each pixel. This allows these coupled diffusion maps or "*CODIMs*" to operate on complete images as input rather than low-dimensional vectors.

2. A Prelude to CODIMs

This section sketches the research that has lead up to the coupled diffusion schemes. The CODIM framework draws on two recent strands of research.

The first is so-called "regularisation". Regularisation amounts to minimising cost functionals that penalise several unwanted effects in the image The enhanced image corresponds to the intensity function that corresponds to the minimum. Consider the following regularising functional for the original image g:

$$E(f, B) = \int_{R-B} [\alpha||\nabla f||^2 + \beta(f - g)^2]dxdy + \nu|B| \qquad (2.1)$$

In this expression α, β and ν are adjustable parameters. One tries to find a piecewise smooth function f – the enhanced image – that on the one hand is close to g (cfr. the $(f - g)^2$ penalising term) but on the other is smooth (cfr. the term $||\nabla f||^2$, i.e. the image gradient squared), except on the set of image edges B (the size of which is measured using some appropriate measure $|B|$). Unfortunately, the functional is non-convex and minimisation therefore is highly non-trivial. Also other problems such as over-segmentation and rigid constraints on the edges plague this approach.

These limitations prompted Shah to propose a different set of minimising functionals (the details of which can be found in [4]):

$$F = \int_R [||\nabla f||^2 + \frac{1}{v^2\sigma^2}(f - g)^2] \, dxdy$$

$$V = \int_R [\alpha(1 - v)^2||\nabla f|| + \frac{\rho}{2}||\nabla v||^2 + \frac{v^2}{2\rho}] \, dxdy$$

Here, the function v is an *edge-indicator* for the grey-level f. It indicates the likelihood of there being an edge at that position in the image: hence, v is assumed to be smooth and close to 1 in the vicinity of an f-discontinuity, and close to zero away from the edges.

Since the functionals F and V are convex, one can search for their minima by equating their first variations δF and δV to zero. This yields the Euler-Lagrange equations for the functionals. Alternatively, one can opt for a more dynamically inspired solution by writing down the evolution equations which one obtains by forcing the time evolution to be proportional but opposite to the first variations:

$$\frac{\partial f}{\partial t} = -\delta F \qquad \frac{\partial v}{\partial t} = -\delta V.$$

This yields (cfr. [4]):

Shah: $\qquad \begin{cases} \dfrac{\partial f}{\partial t} = v^2\nabla^2 f - \dfrac{1}{\sigma^2}(f - g) \\[2mm] \dfrac{\partial v}{\partial t} = \rho\nabla^2 v - \dfrac{v}{\rho} + 2\alpha(1 - v)||\nabla f||. \end{cases}$ $\qquad (2.2)$

The set of first variations above stabilises rather quickly during the iteration process, albeit on a solution that may look noticeably blurred. Notice that the two coupled equations are indeed smoothing processes as they involve the terms $\nabla^2 f$ and $\nabla^2 v$. In fact v can be considered as indicative of a region of interest along the edges. The width of this region is determined by the smoothness coefficient ρ. Moreover, v behaves non-linearly with respect to the contrast (gradient) in the image, and turns out to be more convincing than the normal gradient approach. The smoothing however, results in a loss of localisation of the boundary.

This brings us to the second research track to which the CODIM framework is indebted. Instead of smoothing an image by running a diffusion process, Perona & Malik [2] proposed a method to simultaneously sharpen the edges. This can be achieved by making the diffusion coefficient c in

$$\frac{\partial f}{\partial t} = c\nabla^2 f = c\Delta f$$

dependent on the gradient of f:

$$\text{Perona-Malik:} \qquad \frac{\partial f}{\partial t} = div(c(\|\nabla f\|)\nabla f) \qquad (2.3)$$

Choosing $c = 0$ at the edges and $c = 1$ elsewhere, encourages smoothing within a single region in preference to smoothing across its edges. This is achieved by making c a function of the intensity gradient that falls off to 0 for large gradient magnitudes. The result is that gradients above a certain threshold are enhanced, whereas below this threshold, diffusion will tend to have a smoothing effect. Instead of being too smooth, this scheme yields disturbing "contouring" effects (staircasing). Also there is a problem with discrete implementations in that they will tend to a homogeneous end state.

With eqs. (2.2) leading to non-homogeneous but blurred images and eq. (2.3) to sharp edges but also contouring and homogeneous end states, it seems natural to integrate these schemes to combine their strengths and attenuate their weaknesses, as discussed next.

3. CODIMs for Edge Preserving Denoising

A natural way of introducing the Perona & Malik (P & M) operator (2.3) into the Shah eqs. (2.2) is [3]:

$$\begin{aligned}
\frac{\partial f}{\partial t} &= v^2 div(c(\|\nabla f\|)\nabla f) - \frac{1}{\sigma^2}(f - g) \\
\frac{\partial v}{\partial t} &= \rho\nabla^2 v - \frac{v}{\rho} + 2\alpha(1 - v)\|\nabla f\|
\end{aligned} \qquad (3.1)$$

The first term of the first equation guides the smoothing / edge sharpening P & M diffusion. In fact, removing the v^2 in the first equation turns it into the equation introduced by Nordström [1]. Removing v^2 would further reduce noise away from the edges, but would also cause more contouring (i.e. unwanted edge sharpening). The effect of the $(f - g)$-term is to keep the result sufficiently close to the original image intensity g, thereby keeping some texture and also attenuating the contouring effect. Note that $c(\|\nabla f\|)$ can also be replaced by $c(v)$. The first equation can be considered to yield an enhanced "intensity map" (i.e. enhanced image), whereas the second one evolves to an "edge map", with high values of v at edges in the f-map. Such split of the information is what we referred to as specialised modules in the introduction. Notice that the information transfer between f and v is bi-directional.

This system was used as a preprocessing step in a crop classification problem based on SPOT images. Figure 3.1 shows an original SPOT-satellite image and edges found by the Sobel edge detector. Figure 3.2 shows the processed image (f) and the associated edge map (v). As can be seen, both the intensity and the edges have been improved. The median filter is provided for reasons of comparison.

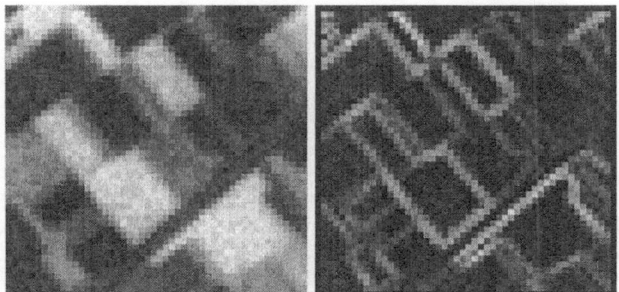

Fig. 3.1. Original SPOT image and its Sobel edge map

Fig. 3.2. Processed SPOT-image and its edge map (left and middle) and the result of a 3×3 median filtering (right).

The system (3.1) applies the restraining force $(f-g)$ to all pixels, including the edges, and thereby counteracts the edge sharpening. If one wants to preserve even small intense spots as clearly visible spikes, one can let the smoothing / edge sharpening P & M operator have its full effect away from the major edges (eliminate v^2) and suppress the restraining force at all edge-like pixels:

$$\frac{\partial f}{\partial t} = div(c(\|\nabla f\|)\nabla f) - (1 - v)\frac{1}{\sigma^2}(f - g)$$

$$\frac{\partial v}{\partial t} = \rho\nabla^2 v - \frac{v}{\rho} + 2\alpha(1 - v)\|\nabla f\| \tag{3.2}$$

Isolated edge pixels can now more easily survive, provided they are of sufficient contrast, e.g. outliers in a field (instead of corrupting the surrounding data). This has also been applied to SAR speckle noise reduction. Figure 3.3 shows the original, noisy image, the result after applying (3.2), and the result of median filtering as a reference. The challenge is to smooth the noise without eroding the structures of interest: in this case, eg the bridge across the river (cf. the narrow stripe). The resulting images have been used to detect coast lines.

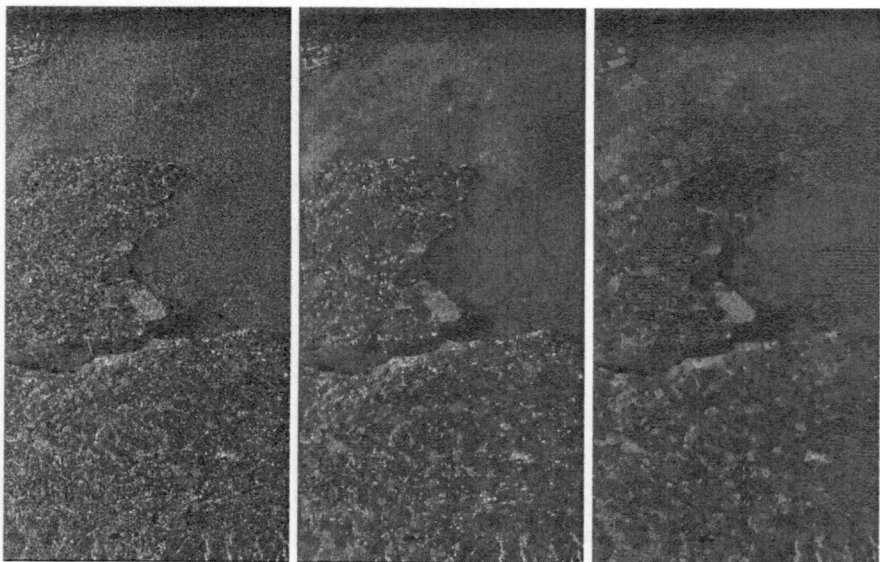

Fig. 3.3. Original SAR-image (left), after diffusion (middle) and median-processed (right) image.

4. Multispectral Images

The idea of non-linear diffusion is not restricted to one intensity maps, but can easily be extended for multi-spectral applications. For example, suppose there are three channels – red, green, and near infrared, say – that have to be preprocessed. One way would be to process them independently, but one map (channel) can benefit from the information in the others. If one map shows a high gradient at some point, this gradient should also influence diffusion in the corresponding points in the other maps. Figure 4.1 shows a synthetic set of three noisy images g_r, g_g and g_n. Each channel shows one bright patch, and three dark patches, one of which has a slightly different intensity.

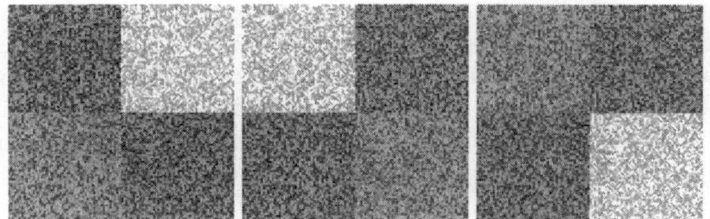

Fig. 4.1. Synthetic image components g_r, g_g, g_n.

One can use the system:

$$\frac{\partial f_r}{\partial t} = div(c(\beta, \tau)\nabla f_r) - \frac{1-v}{\sigma^2}(f_r - g_r)$$

$$\frac{\partial f_g}{\partial t} = div(c(\beta, \tau)\nabla f_g) - \frac{1-v}{\sigma^2}(f_g - g_g)$$

$$\frac{\partial f_n}{\partial t} = div(c(\beta, \tau)\nabla f_n) - \frac{1-v}{\sigma^2}(f_n - g_n)$$

$$\frac{\partial \beta}{\partial t} = \xi\nabla^2\beta - (\beta - max(\nabla f_r, \nabla f_g, \nabla f_n))$$

$$\frac{\partial v}{\partial t} = \rho\nabla^2 v - \frac{v}{\rho} + 2\alpha(1 - v)\|\beta\|$$

with $c(\beta, \tau) = \frac{1}{1+(\beta/\tau)^2}$ where τ is a constant parameter. The three channels, governed by the first three equations, interact through the latter two. Their effect is that diffusion is stopped across edges where the maximum gradient of the three channels becomes large, irrespective of how small the other two gradients may be. In the example, this set of equations will clearly sharpen all the discontinuities (figure 4.2), since for each of the four discontinuities, there is at least one channel with a high gradient.

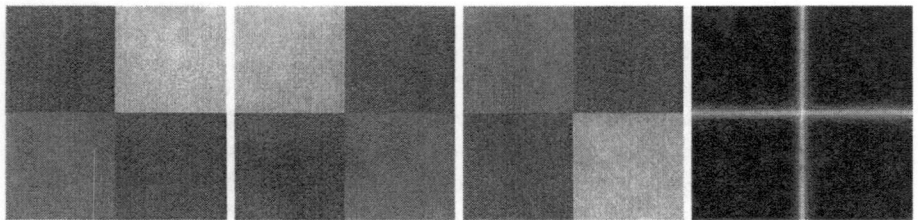

Fig. 4.2. Multi-spectral approach f_r, f_g, f_n, v ($\tau = 10.0, \sigma = 7.0, \xi = 0.0, \rho = 3.0, \alpha = 0.05$)

5. Crop Classification as an Example

Several preprocessing operations were compared, that each reduce noise in multi-spectral SPOT data used to distinguish 7 crops. All preprocessed data were combined with exactly the same CART classifier. Figure 5.1 shows the classification rates for several averaging, median, and two non-linear diffusion schemes. As expected, the linear averaging ("mean") performs worse than the

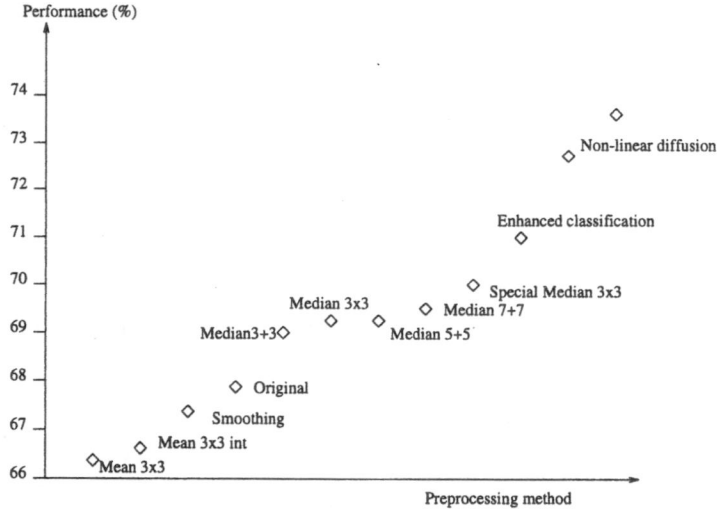

Fig. 5.1. Crop classification rates with different types of preprocessing.

non-linear schemes (or without preprocessing!). However, the two non-linear diffusion schemes still clearly add to the quality compared to results obtained with median filtering. The lowest scoring diffusion scheme corresponds to a multi-spectral version of eqs. (3.1), the highest scoring to the multi-spectral implementation for eqs. (3.2) as previously given. The former might give perceptually nicer results, but that is of less importance here.

6. Conclusion

CODIMs exhibit many attractive characteristics for low-level processing: they are highly parallelizable and are therefore natural candidates for VLSI-implementation. Their coupling ensures consistency and mutual support between different features that are each interesting in their own right (e.g. intensities and edges). The non-linearity allows for a rich and interesting dynamics.

Application of CODIM schemes to optical flow, stereo reconstruction, etc. are given elsewhere [3]. In the meantime, CODIMs have also been developed for texture segmentation and the detection of mirror symmetries and repeated patterns. Another part of the work has been to develop VLSI for the Nordström equation. The corresponding chip is expected to be ready soon and a PCI board with several chips will be built to allow for video rate application of this equation (e.g. 100 iterations in $40ms$ on a 256×256 image.

Acknowledgement. The support by Esprit-LTR 'IMPACT' is gratefully acknowledged. MP also gratefully acknowledges support by a Post-Doctoral grant of the Flemish Inst. for the Promotion of Scient.-Techn. Research in Industry.

References

1. N. Nordström, "Biased Anisotropic Diffusion: A Unified Regularisation and Diffusion Approach to Edge Detection", *Image and Vision Computing*, vol. 8, no. 4. 1990.
2. P. Perona and J. Malik, "Scale-Space and Edge Detection Using Anisotropic Diffusion", *IEEE Transactions on Pattern Analysis and Machine Intelligence*, vol. 12, no. 7, 1990.
3. M. Proesmans, E. Pauwels, and L. Van Gool, "Coupled geometry-driven diffusion equations for low-level vision", in: *Geometry-driven Diffusion in Computer Vision*, ed. B. ter Haar Romeny, pp. 191–228, Kluwer, 1994.
4. J. Shah, "Segmentation by non-linear diffusion", in: *Proceedings IEEE Conference on Computer Vision and Pattern Recognition* Hawai, 1991.

Application of the Constructive Mikado-Algorithm on Remotely Sensed Data

C. Cruse[1], S. Leppelmann[2], A. Burwick[2], and M. Bode[3]

[1] Institut für Geoinformatik, Universität Münster
[2] Institut für Angewandte Physik, Universität Münster
[3] Max-Planck-Institut für Physik komplexer Systeme, Dresden.
   e-mail: cruse@uni-muenster.de

Summary. Finding an optimal architecture for a neural network, i.e. an optimal number and size of layers, is an open problem. With the Mikado-algorithm we present a new method to construct the network architecture in the course of learning. With two examples, classification and structure detection from remotely sensed data, we demonstrate the capabilities of the Mikado-algorithm. This algorithm provides good generalisation in the presence of mixed pixels, delivering small networks for high-dimensional problems and a new way of interpreting the network generalisation ability.

1. Introduction

The main problem in remotely sensed image classification is the generation of a classifier which allows both a fast calculation and good generalisation. We present the Mikado-algorithm which provides a new solution for developing small-sized networks especially in the case of high dimensional data. The Mikado-algorithm [1, 2] is a constructive algorithm which is characterised by a fully correct representation of the training set, fast learning and a good generalisation ability. High dimensional data often carries a significant amount of redundant information. Thus it seems useful to reduce the dimension of the input representation before applying the actual classification algorithm. Our approach is to reduce the input dimension of the neurons, i.e. the connectivity of these neurons, in a systematic way. In particular, computational speed profits from this reduction and attempts to find a conceptual interpretation of the classification process are simplified.

Section 2 introduces the fundamental principles of the Mikado algorithm. After this the classification ability of the Mikado-algorithm on satellite images is examined and it is demonstrated how mixed pixels can be treated successfully. Section 4 gives an example for the automatic detection of noisy structures, i.e. blind shells (or unexploded-bomb impact-craters), in aerial images. In section 5 the capability to reduce the net size is discussed. Finally the main aspects of the Mikado are summarised.

2. The Mikado-Algorithm

The Construction of the first layer by the Mikado is founded on the principle of faithfulness [3]. Faithfulness means, that two training patterns of different classes are mapped onto different outputs. If this is not the case we have a contradiction and a hyperplane (threshold neuron) is introduced to separate these patterns (figure 2.1). This procedure is to be applied until no contradiction is found for all patterns. This coding layer can be interpreted as preprocessing stage. A modified LVQ3 algorithm [4] is used to construct the next layer (address layer). The last layer only implements a 1 of n coding (see table 2.1). The scheme for the design of the coding layer is given by the following steps:

> 1. Find two patterns having a contradiction
>   if no contradictions found ⇒ end
>   else goto step 2.
> 2. Separate patterns by adding a hyperplane (threshold neuron).
> 3. Return to step 1.

For the calculation of weight vector w and the threshold of a neuron, the mean value construction scheme is introduced (figure 2.1). The mean value construction introduces a orthogonal-hyperplane centred on the straight line between two patterns of different classes. These hyperplanes separate the state space into different sectors.

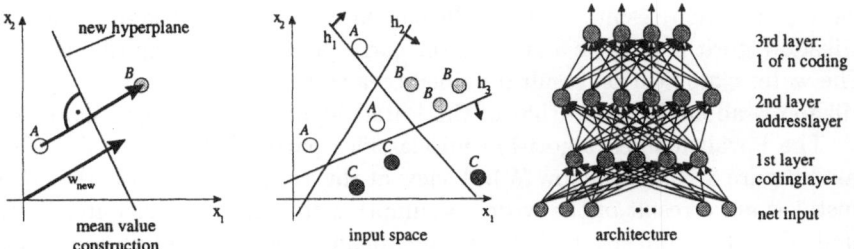

Fig. 2.1. Construction of the first layer. Left: Mean value construction principle of the hyperplane to separate two patterns of classes A and B in the 2 dimensional input space X. Middle: Structuring of the input space X given by the neurons $h1$, $h2$, $h3$. The arrows show the direction of the weight vectors. Right: Architecture of the network.

At the the end of the construction of the first layer these sectors contain one or more patterns of only a single class. After all hyperplanes are constructed the threshold neurons are replaced by sigmoid neurons.

The construction of the second layer (address layer) is done by a constructive vector quantisation. This layer is trained by the LVQ3 algorithm [3]. We

Table 2.1. The table shows the neuron types and training algorithms for the Mikado vector quantisation networks.

layer	neuron type for training	algorithm for training	neuron type for evaluation
coding layer	threshold	Mikado	sigmoid neuron
address layer	codebook vector	LVQ3	radial basis function
1 of n coding	sum of inputs	-	winner takes all

start with a small number of neurons and apply the LVQ3 for a certain number of epochs. As long as the error is not vanishing, new neurons are added [4]. This combination of training and construction has to be continued until the desired quality of classification is reached. After all neurons are constructed they are replaced by radial basis functions. The last layer performs a 1 of n coding (figure 2.1 right). The recognised class is determined by a winner takes all evaluation.

3. The Mikado-Algorithm for Classifying Remotely Sensed Images

As a first experiment we compare the Mikado with other classifiers using data from Landsat Thematic Mapper (LTM) images of the town of Münster in the north west of Germany. The classification task was to detect areas of water in the city. To train the network 15 examples for water and 74 examples of vegetation and urban area were taken. The following four images show the classification results of a Parallelepiped algorithm [5] (figure 3.1a), the Mikado-algorithm (figure 3.1b) and the corresponding city map (figure 3.1c). The water classification result obtained by a comparison with the map of the Mikado is about 98% and that of the Parallelepiped 81% recognition rate.

The Parallelepiped algorithm misclassifies parts of the lower of the two lakes (figure 3.1b; 3.1c arrow W). A view at the scattergram shows that these mistakes are a result of the wrong assumption that the data of water is normally distributed in spectral space. The Mikado does not use this assumption and so gets superior results.

Furthermore one can see that by the Parallelepiped algorithm some parts of water around the castle are not classified (figure 3.1b; 3.1c arrow C). The problem here is that most of the $30m \times 30m$ areas do not consist of water but of trees. One possibility to make the water structure visible is to calculate the probability function of water [6]. This is done by using not only one Mikado-network but an ensemble of networks. In contrast to statistical methods such as the Parallelepiped or the Maximum Likelihood classifiers each Mikado net classifies each example pattern correctly but nevertheless generalises a little differently. By evaluation of 100 different nets we obtained figure 3.1d) which shows the number of nets on each pixel which classify water. Now structures

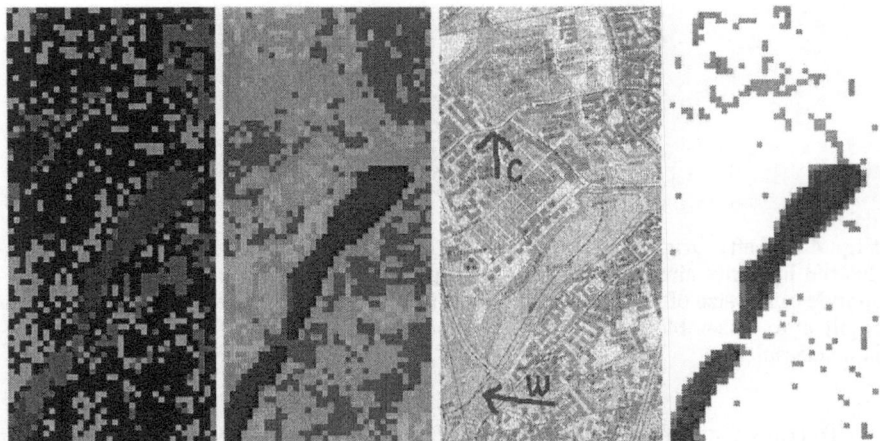

Fig. 3.1. a) Classification by a Parallelepiped classifier, red: urban area, blue: water,green vegetation, black: not classified; b) Classification by the Mikado classifier c) Aspect of the city map of Münster d) The probability function of water shows structures in the upper part of the image, that is the water around the castle garden.

of the water around the castle garden are visible, that could not be seen before.

4. The Mikado-Algorithm for Structure Detection

This section will demonstrate that the Mikado is an efficient classification algorithm for image processing. As an example we use the automatic detection of blind shells in aerial images taken in World War II. It is an important task nowadays to sweep the blind shells from areas which have to be cultivated. The automatic detection of blind shells on these images will be done in two steps. First the blind shell classifier has to be constructed and then the classifier has to be applied to every pixel of the image.

As opposed to Drewniok [7] the classifier is not calculated from heuristic assumptions but automatically generated by only 58 examples (12 blind shells and 46 blind shell free areas) out of the $\sim$ 30000 pixels of the image, which were identified by experts. The size of the examples is 30×30 pixels, corresponding to the size of the largest blind shell (figure 4.1 left). Figure 4.1 middle shows the classification result of the Mikado which is constructed by an ensemble of 20 networks. The brighter the pixel the higher is the membership to the blind shell class. The four blind shells and the blind shell free area are classified correctly. To get a binary image to decide if a group of pixels is a blind shell location or not, it is easy to introduce a threshold (figure 4.1 middle).

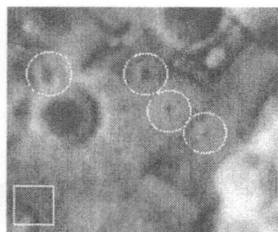 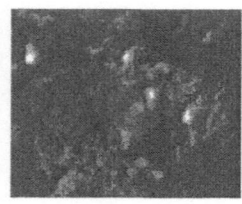 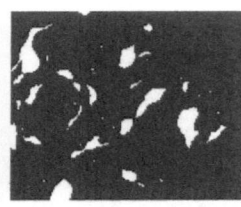

Fig. 4.1. Left: Original aerial image (197 × 171) containing four blind shells (indicated by white circles), identified by experts. The white frame bottom left corresponds to the size of the mask of the classifier (30 × 30 pixel). Middle: Classification result of an ensemble of 20 Mikado networks. Right: Result of the k-nearest neighbour classifier.

To compare this result with other classifiers we used a k-nearest neighbour classifier (k-NN) and tested different k on the same examples as we used with the Mikado. The best classification was obtained with $k = 1$ (figure 4.1 right) and is worse than the Mikado generalisation.

5. The Mikado-Algorithm for Reduction of Connectivity

To reduce the dimension of the input neurons a special variant of the Mikado is used. The network is constructed in two steps. First the difference vector of two patterns of two classes is calculated. The new idea is that all components of this vector except c the maximal of the absolute value are set to zero. This vector is used as the weight vector of the new neuron of the first layer. Each neuron in this layer has only c inputs. The second layer is constructed as before.

5.1 A Simple Example for the Reduction of the Network Size

To motivate our approach and to show the effect of connectivity reduction on the network size and the generalisation we use a training set of 500 handwritten digits. The patterns are scaled to a 16×16 matrix. The aspect ratio is retained and the digits were centred. The resolution is eight bit and the grey values are normalised.

In figure 5.1 left some numbers are shown. In the right figure the weight values of nine neurons of the first layer obtained by the difference between two patterns are presented. One can see two overlapping numbers in each of the 9 images. For discriminating the numbers the extreme white and black pixels are especially relevant. So most of the weights can be deleted without loss of information. In figure 5.2 top left, the resulting number of weights is depicted as a function of the number of inputs per neuron c.

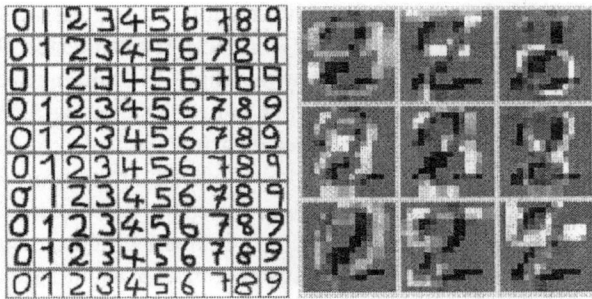

Fig. 5.1. Left: part of the training set of handwritten digits. Right: weights of 9 neurons of the first layer.

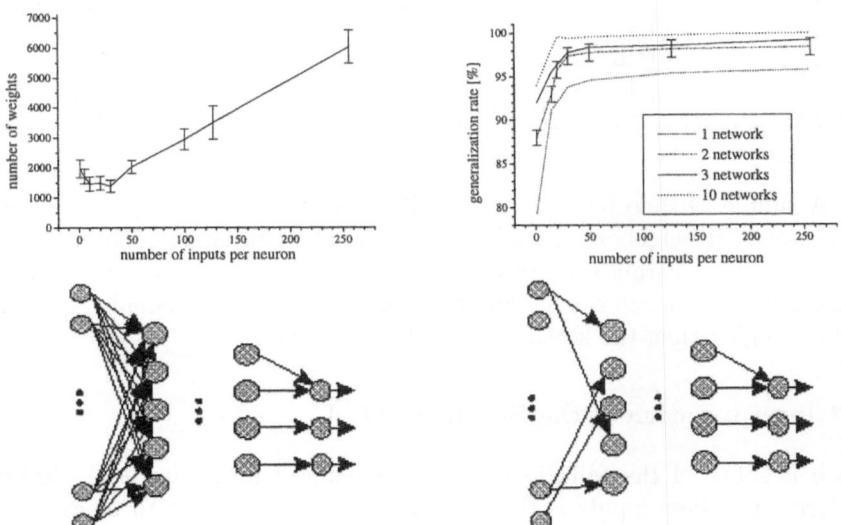

Fig. 5.2. Top left: number of weights of the whole network. Bottom left: rate of correct classified patterns of the test set. Top right: full network architecture. Bottom right: reduced network architecture.

The number of weights of the whole network is minimal at a number of 10 to 30 inputs per neuron. This selection of inputs leads to a classifier which is about four times smaller than using the whole input.

Figure 5.2 bottom left shows the examination of the generalisation ability for a test set of 500 patterns. Without connectivity reduction 95.8% of the 500 patterns are correctly classified by a single Mikado network. For the smallest resulting networks (30 inputs per neuron) the generalisation ability of a single net is 94% that is a misclassification rate of 6%. These results can be improved significantly, if an ensemble of nets is used for classification. The outputs of these networks are first summed up for each class and then the maximum of these summed up outputs is determined as the recognised class. Figure 5.2 bottom left shows that the generalisation rate increases by more than 3% for the classification with two networks. That means that half of the mistakes are gone. More networks lead to even better results. A good compromise between generalisation ability and the size of the classifier seems to be to use two networks. To give an overview, table 5.1 shows some selected networks obtained with the demonstrated reduction methods.

Table 5.1. The network size and generalisation ability for the test set containing 500 patterns for different network types. 1-NN: nearest neighbour classifier, M: Mikado-vector quantisation, M red.: Mikado with reduced connectivity per neuron (15 inputs per neuron).

Type of Network	number of nets	total number of weights	generalisation
k-NN	1	128500	95.8%
M	10	59460	99.6%
M	1	5946	95.8%
M red.	2	3004	97.4%

A simple Mikado net which uses all inputs has the same generalisation as a k-nearest neighbour classifier (k=1) [8]. The evaluation with more nets increases the generalisation to over 99%. The networks with reduced input connectivity are much smaller and the number of correctly classified patterns is even higher than the k-nearest neighbour classifier.

5.2 Interpretation of the Ensemble Mask

Each network of the Mikado uses only a reduced input and the selection criterion for these inputs is the correct classification. It has to be expected that the structure of the object to be classified can be seen in the selected inputs of an ensemble of networks. The resulting mask of this ensemble with all selected inputs is introduced as the ensemble mask.

To get a first impression of the ensemble mask we investigated a one dimensional edge detector with a length of 30 pixels. We take only one exactly centred example for the edge (figure 5.3d, the example in the middle) and 28 counter examples. This training set enables the sharp detection of the centre of the edge in all cases. With these training patterns we constructed 100 networks with maximal reduced connectivity ($c = 1$). We did the same with different disturbed edges, an additional step (figure 5.3b) and a linearised step (figure 5.3c) to compare the constructed ensemble masks.

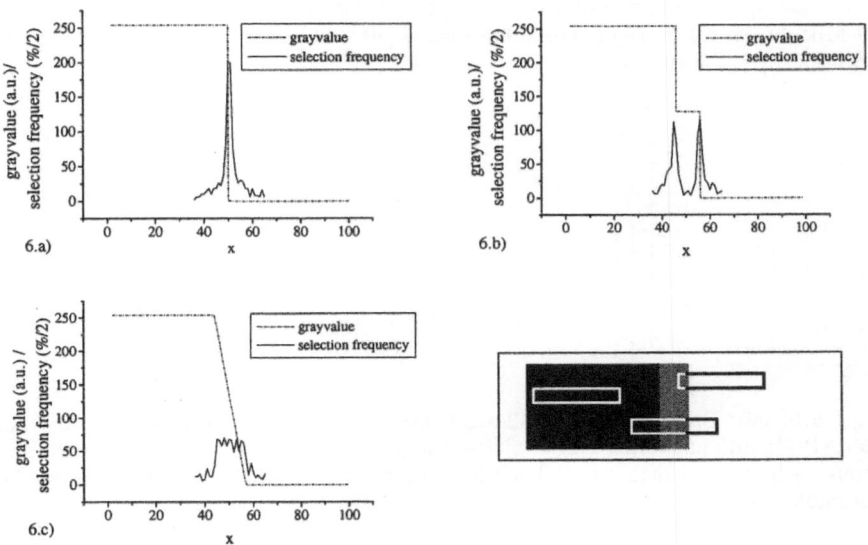

Fig. 5.3. The filter profiles for one-dimensional edge detection.

In figure 5.3 the gray values of the images (dashed line), the corresponding ensemble masks (straight line) and one of the images which has to be classified (bottom right) are shown. The ensemble masks differ according to the kind of shape of the edge. The linearised edge gives an area of high selection probability (figure 5.3c) while an additional edge gives an additional peak according to the distance of the transition (figure 5.3b). The results can easily be understood by remembering that the selection of pixels is done according to the maximal difference of two patterns of different classes. To extend these results to a two dimensional mask the task is now to detect edges in an image with a white square on a black background (figure 5.4 left). The corners were not in the training set. The resulting classification (figure 5.4 middle) and the ensemble mask (figure 5.4 right) are easy to understand as a superposition of two orthogonal one dimensional edge detectors.

To interpret the ensemble mask of the blind shells (figure 5.5 left) we calculated a normal radial selection probability (NRSP) from the ensemble mask. The ensemble mask itself is generated from 400 networks and we

Fig. 5.4. left: image for the edge detection task. The gray frame shows the size of a training pattern. middle: Resulting classification image with an ensemble of 20 Mikado networks. right: ensemble masks (20 × 20). The gray scale encode the selection probability.

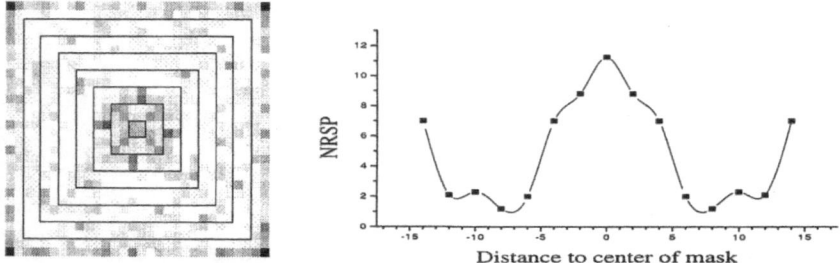

Fig. 5.5. left: Ensemble mask (30 × 30) for blind shell detection generated from 400 AP-Mikado networks. Notice that each network uses only 12 neurons with one input each. right: Normal radial selection probability (NRSP) calculated from the ensemble mask.

summed up all intensities of pixels in each marked frame (see figure 5.5 left) and divided by the number of pixels in that frame.

The result can be seen in figure 5.5 right. The maximum in the centre can be explained by looking at the blind shells inside the circles in figure 4.1 left. There the blind shells have a characteristic change of grey values and according to figure 5.3c this should result in a high selection probability in the centre of the ensemble mask. This was detected by the Mikado without any of the pre-knowledge of the experts. Comparing the synthetic figures with the blind shells, the increase at the edge of the ensemble mask seems not to be due to the blind shells. This could be caused by the effects of the surrounding. Summarising, it can be said that with the ensemble mask one has a new way to get a hint for the generalisation ability of the network ensemble.

6. Summary

In this work the Mikado-algorithm was introduced as a powerful method for automatic construction of small classifiers with high generalisation ability in a systematic way.

The Mikado constructed networks used in our applications were one order of magnitude smaller in net size compared to a k-nearest neighbour classifier.

It was shown for classification of remotely sensed images that the mixed pixel problem can be solved and that noisy structures in the image can be found. Furthermore the input channels can be evaluated for their use in classification. This can be seen as a first step to a plausibility analysis of the constructed neural network.

References

1. C. Cruse, M. Bode, and H. G. Purwins, Ein Konstruktionsalgorithmus für Mustererkennungssysteme, in: Physik und Informatik - Informatik und Physik, Arbeitsgespräch München, 21-22 November 1991, Proceedings (Eds.: D. Krönig and M. Lang), Springer-Verlag, 1991.
2. C. Cruse, Konzepte der Effizienz in künstlichen Neuronalen Netzwerken, Doktorarbeit, Universität Münster, 1996.
3. M. Mezard, J. P. Nadal, "Learning in Feedforward Layered Networks: The Tiling Algorithm", *Journal of Physics A*, vol. 22, pp. 2191–2204, 1989.
4. T. Kohonen, J. Kangas, J. Laaksonen, and K. Torkkola, LVQ-PAK, The Learning Vector Quantization Program Package, Version 2.1, LVQ-Programming Team of the Helsinki University of Technology, Lab. of Comp. and Inform. Science, Rakentajanaukio 2 C, SF-02150 Espoo, 1992.
5. T. M. Lillesand, and R. W. Kiefer, *Remote Sensing and Image Interpretation*, 2nd Edition, 1987.
6. T. Vieth, "Möglichkeiten zur Verbesserung von satellitengestützen Landnutzungsklassifikationen durch Verknüpfung mit topographischen Informationen am Beispiel des Einzugsgebietes Halterner Stausee", Diplomarbeit, Institut für Geoinformatik, Universität Münster, 1995.
7. C. Drewniok, and K. Rohr, "High precision localization of circular landmarks in aerial images", in: *Mustererkennung 1995, 17. DAGM-Symposium, Bielefeld, September 1995*, Springer Verlag, 1995.
8. M. Neschen, "Hierarchical Binary Vector Quantisation Classifiers for Handwritten Character Recognition, Zentrum für Paralleles Rechnen (ZPR)", Universität zu Köln, D-50931 Köln, Germany, in: *Mustererkennung 1995, 17. DAGM-Symposium, Bielefeld, September 1995*, Springer Verlag, 1995.

A Simple Neural Network Contextual Classifier

Jens Tidemann and Allan Aasbjerg Nielsen

IMM – Department of Mathematical Modelling,
Technical University of Denmark, Building 321,
DK–2800 Lyngby, Denmark.
http://www.imm.dtu.dk

Summary. In this paper we describe a neural network used to make a simple contextual classifier using a two layer feed-forward network. The best number of hidden units is chosen by training a network with too many hidden units. We then prune the network using Optimal Brain Damage (OBD). The pruned networks have a better generalisation error because they only have the weights that reflect the structure of the data and not the noise. We study the possibility of using a Network Information Criterion (NIC) to decide when to stop pruning. When we use NIC we can estimate the test error of a network without using an independent validation set.

As a case study we use a four band Landsat-2 Multispectral Scanner (MSS) image from southern Greenland. To classify a pixel in the non-contextual case we use the four variables from the MSS bands only. In the simple contextual case we augment the feature vector with the four mean values of the MSS bands from the four nearest neighbours. We notice an increase in the number of correct classified pixels when using the contextual classifier. Also, the application of the simple contextual classifier gives a small overall increase in the posterior probability.

1. Introduction

In this paper we study the use of neural networks for classification of remote sensing images. When classifying remote sensing data it is important to use contextual information because the data often have a lot of noise.

When working with neural networks it is important to find the correct network size. We study the use of pruning weights in the network. Thereby we can train a network that is too big, prune some of the connections and then choose the network that is optimal. We investigate the use of a Network Information Criterion to find the optimal network.

In section 2 we describe the 2-layered network architecture that we use and in section 3 we discuss the optimisation method. In section 4 we discuss Optimal Brain Damage, the method that we use to remove weights from the network. Section 5 deals with a Network Information Criterion that can be used to test which of the many networks we train is best. In section 6 we discuss the contextual classification and section 7 is devoted to a case study where we apply our methods to an image taken with the Landsat-2 Multispectral Scanner (MSS).

2. Network Architecture

We use a 2-layer feed forward network. In the hidden layer we use hyperbolic tangent as activation function. The weight from input i to hidden unit j is denoted as w_{ji}. We include a bias by setting $x_0 = 1$. So a hidden unit is described by the equations

$$a_j = \sum_i w_{ji} x_i \quad , \ x_0 = 1 \tag{2.1}$$

$$z_j = \tanh(a_j) \tag{2.2}$$

In the output layer we use soft-max as activation function (a_k is found by weighting the output from the hidden layer z_j, like (2.1))

$$y_k = \frac{\exp(a_k)}{\sum_{k'} \exp(a_{k'})} \tag{2.3}$$

We use the cross-entropy as energy function

$$E = \frac{1}{N} \sum_{n=1}^{N} \sum_{k=1}^{c} t_k^n \ln\left(\frac{y_k^n}{t_k^n}\right) \tag{2.4}$$

where N is the number of training samples, c is the number of classes and t_k^n is the target value for observation n in class k.

By choosing soft-max as activation function in the output layer and cross-entropy as energy function we have ensured that the network is optimal for classification and we can interpret the output as probabilities [1].

3. Network Training

We use a Broyden-Fletcher-Goldfarb-Shanno (BFGS) quasi Newton method [4] for optimising the network. In a Newton method you make a second order estimation of the energy function and make a step towards the minimum

$$\mathbf{w}^{\tau+1} = \mathbf{w}^{\tau} + \alpha^{\tau} \mathbf{G}^{\tau} \mathbf{g}^{\tau} \tag{3.1}$$

The α parameter is used to make sure that the energy is decreasing. It is found by line minimisation. When using a second order method you have to calculate the second derivative of the energy function (the Hessian matrix). This gives two problems. The Hessian might not be positive definite and is computationally expensive to estimate. Instead we use a quasi Newton method. This is much faster and we can ensure that the estimate of the Hessian will always be positive definite. We update the estimate of the Hessian after each step. The updating equation for the estimate of the Hessian ($\mathbf{G}^{\tau}$) is

$$\mathbf{G}^{\tau+1} = \mathbf{G}^\tau + \frac{\mathbf{p}\mathbf{p}^T}{\mathbf{p}^T\mathbf{v}} - \frac{(\mathbf{G}^\tau\mathbf{v})(\mathbf{v}^T\mathbf{G}^\tau)}{\mathbf{v}^T\mathbf{G}^\tau\mathbf{v}} + (\mathbf{v}^T\mathbf{G}^\tau\mathbf{v})\mathbf{u}\mathbf{u}^T \qquad (3.2)$$

We have defined these vectors:

$$\mathbf{p} = \mathbf{w}^{\tau+1} - \mathbf{w}^\tau \qquad (3.3)$$

$$\mathbf{v} = \mathbf{g}^{\tau+1} - \mathbf{g}^\tau \qquad (3.4)$$

$$\mathbf{u} = \frac{\mathbf{p}}{\mathbf{p}^T\mathbf{v}} - \frac{\mathbf{G}^\tau\mathbf{v}}{\mathbf{v}^T\mathbf{G}^\tau\mathbf{v}} \qquad (3.5)$$

Where $\mathbf{w}$ are the weights and $\mathbf{g}$ is the gradient. When we have estimated the Hessian we make a Newton step (3.1) in the descent direction of the energy function. In the line minimisation we make a parabolic fit [4] to the energy function. The algorithm is initiated by setting $\mathbf{G}^0$ to the identity matrix, so the first step is a gradient descent step.

4. Optimal Brain Damage

When we start training the network we have a fully connected network. But it is not certain that all the connections are needed for the classification. By pruning the network we increase the generalisation of the network because we reduce the number of parameters in the model without increasing the energy function significantly. The idea in Optimal Brain Damage (OBD) [3] is to estimate the increase in the energy function when we remove one weight from the network (the saliency for that weight). We then remove the weights with the lowest saliencies and retrain the network. We continue this process until the energy function starts growing dramatically.

OBD can be used to decide how many hidden units will be needed in the network. We can train a network that we know is too big and then we can prune it. During the pruning some of the hidden units will be removed. We can also use it to decide if any of the features we use in the network are unimportant. If a feature does not contribute to the classification all weights for that feature will be removed from the network.

When we estimate the increase in energy we make a Taylor series for the change in the energy. We assume that we are in a local minimum so the gradient is $\mathbf{0}$. Instead of calculating the full Hessian matrix we assume that it is diagonal and we make a diagonal approximation to the Hessian

$$\begin{aligned} \delta E &= \frac{1}{2}\delta\mathbf{w}^T\mathbf{H}\delta\mathbf{w} + O(\|\delta\mathbf{w}\|^3) \\ &\approx \frac{1}{2}\sum_i \delta w_i^2 H_{ii} \end{aligned} \qquad (4.1)$$

When we calculate the diagonal approximation to the Hessian we further assume that off-diagonal elements have no influence on the diagonal elements

$$\frac{\partial^2 E^n}{\partial w_{ji}^2} = \frac{\partial^2 E^n}{\partial a_i^2} x_j^2 \tag{4.2}$$

$$\frac{\partial^2 E^n}{\partial a_j^2} = g'(a_j)^2 \sum_k w_{kj}^2 \frac{\partial^2 E^n}{\partial a_k^2} + g''(a_j) \sum_k w_{kj} \frac{\partial E^n}{\partial a_k} \tag{4.3}$$

By making this assumption we can reduce the computational time dramatically.

5. Network Information Criterion

The Network Information Criterion (NIC) [5] estimates the test energy for a neural network without using a test dataset. So when training and pruning a neural network you do not need an independent validation dataset to select the best network. When you calculate the NIC you use your energy function (2.4) and add a term for the complexity of the model

$$\text{NIC} = E(\mathbf{x}, \mathbf{t}, \mathbf{w}) + \frac{1}{N} tr(\mathbf{G}(\mathbf{w})\mathbf{Q}(\mathbf{w})^{-1}) \tag{5.1}$$

where $\mathbf{x}$ is a matrix with all input data for the network and $\mathbf{t}$ is a matrix with all the target values. We have defined

$$\mathbf{G}(\mathbf{w}) = V[\nabla e(\mathbf{x}^n, \mathbf{t}^n, \mathbf{w})] \tag{5.2}$$

$$\mathbf{Q}(\mathbf{w}) = E[\nabla\nabla e(\mathbf{x}^n, \mathbf{t}^n, \mathbf{w})] \tag{5.3}$$

where e is the energy for one observation from the dataset and ∇ means the gradient.

Murata [5] argues that the models you compare with NIC should be hierarchical, so that each model is a sub-model of the other models. But Ripley [6] proves that this is not necessary. It is discussed in [6] how you should estimate (5.2) and (5.3). If NIC should work properly it requires that there is a strong single local minimum. If this is not the case the complexity term might change dramatically depending on which local minimum you end up in. Another assumption is that there is enough data in the training set.

6. Contextual Classification

We wish to use contextual information in the classification. We could use the features from all the neighbouring pixels. But that would make the number of weights 5 times bigger in the input layer. Instead we assume in our model that the pixel on the north side of the pixel has the same influence on the classification as the pixel on the east, south and west side. We can do this by forcing the weights in the input layer to be equal for the features from the neighbouring pixels.

But we can find a simpler approach if we look at equation (2.1). If we force some of the weights to be equal we get

$$
\begin{aligned}
a_j &= \cdots + w x_N + w x_E + w x_S + w x_W + \cdots \\
&= \cdots + w(x_N + x_E + x_S + x_W) + \cdots
\end{aligned}
\tag{6.1}
$$

From this we see that instead of forcing the network to have equal weights we sum the features that have equal weights before we start training the network.

In the case of four neighbours we get the Switzer-filter [7] except for a scaling factor. In the Switzer-filter you take the mean value of the neighbouring pixels

$$
y = \frac{1}{4}(x_N + x_E + x_S + x_W)
\tag{6.2}
$$

So we can introduce contextual information to the network by using the Switzer-filter. We thereby avoid a huge increase in network size.

7. A Case Study

As a case study we use a 512×512 pixel four band Landsat-2 Multispectral Scanner (MSS) image from southern Greenland. In Figure 7.1 is shown two of the bands of the image. We have five different training classes in the image with a total of 43000 pixels. We split the dataset up so $\frac{3}{4}$ of the pixels are used for training and the last $\frac{1}{4}$ are used for testing.

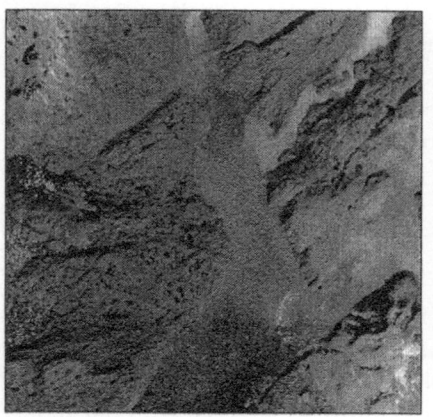

Fig. 7.1. Bands 4 and 7 of the Landsat MSS data.

We first study the NIC. In Figure 7.2 we have plotted the training- and test energy and the NIC against the number of weights in the network during one training and pruning session. We notice that the NIC is very unstable when there are many weights in the network. This might be because the number of weights is high compared with the number of data samples. So we have to be careful when we use NIC particularly when we have many weights in the network. But for a smaller number of weights it seems to be useful. When we prune the network we are able to remove half of the weights in the network and as can be seen from Table 7.1 we get a small increase in the generalisation because the pruned networks perform better. We train ten networks and prune them. We choose the network with the lowest NIC if the NIC is stable.

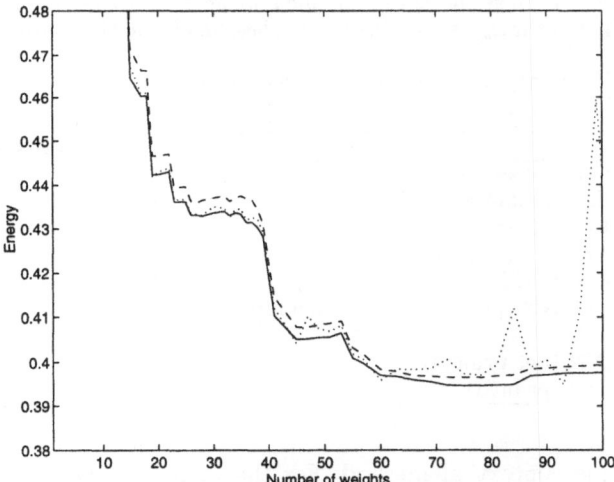

Fig. 7.2. The training and test energy and NIC during one training and pruning session. The full line is the training error, the dashed line is the test error and the dotted line is the NIC.

In Figure 7.3 we see the classified images. It is seen that in the image where we use the contextual classifier we get a significant reduction in the number of single pixels that are classified into another class than their neighbouring pixels. If we study Table 7.1 we see how many pixels are classified correctly for each of the classifiers. The classification of all classes is improved when we use the contextual classifier. When using the contextual classifier there is an overall increase in the probability which means that the network is more certain that it has made the correct classification.

We have compared our results with other methods. If we use linear or quadratic discriminant analysis [6, 1] we get results that are not as good as the non-contextual neural network. If we use CART [2] we get results that are comparable with neural networks. We have also compared our results with

Fig. 7.3. The left image shows the classified image when we use the non-contextual classifier. The right image shows the image classified using the contextual classifier.

Table 7.1. Percentage of pixels that are classified correctly in the test set with the methods we have used.

Method	Correctly classified
Linear discriminant analysis	78.5 %
Quadratic discriminant analysis	79.7 %
Non-contextual NN (not pruned)	81.6 %
Non-contextual NN (pruned)	81.8 %
Non-contextual CART	81.8 %
Non-contextual NN (pruned) and modus filtering	84.8 %
Contextual CART	85.1 %
Contextual NN (not pruned)	86.6 %
Contextual NN (pruned)	86.8 %

another simple contextual method. For the image classified with the non-contextual classifier we have made a modus filtering of the classified image with a 3×3 kernel. The modus filter counts how many pixels belong to each class in the kernel and then assigns the most frequent class to the centre pixel. The result is shown in Table 7.1. It is better than the non-contextual classifier but not as good as the classifier where we use the contextual information as features to the neural network.

8. Conclusion

We have studied pruning of the network weights and we have come to the result that approximately half the weights can be removed from the network and this reduction results in an increase in generalisation error. The use of simple contextual information can improve the classification significantly. Also we get the best results if we use contextual information as input for the neural networks instead of doing post-processing on the data.

References

1. C. M. Bishop, *Neural Networks for Pattern Recognition*. Oxford University Press, 1995.
2. L. Breiman, J. H. Friedman, R. A. Olshen, and C. J. Stone, *Classification and Regression Trees*. Wadsworth and Brooks/Cole, 1984.
3. Y. Le Cun, J. S. Denker, and S. A. Solla, "Optimal brain damage", *Advances in Neural Information Processing Systems*, vol. 2, pp. 598–605, 1990.
4. D. G. Luenberger, *Linear and Nonlinear Programming*, Addison-Wesley, 2nd edition, 1984.
5. N. Murata, S. Yoshizawa, and S. Amari, "Network information criterion - determining the number of hidden units for an artificial neural network model", *IEEE-Transactions on Neural Networks*, vol. 5, no. 6, pp. 865–872, 1994.
6. B. D. Ripley, *Pattern Recognition and Neural Networks*, Cambridge University Press, 1996.
7. P. Switzer, "Extensions of linear discriminant analysis for statistical classification of remotely sensed satellite imagery", *Mathematical Geology*, vol. 12, no. 4, pp. 367–376, 1980.

Optimising Neural Networks for Land Use Classification*

Horst Bischof and Aleš Leonardis

Department for Pattern Recognition and Image Processing,
Technical University Vienna,
Treitlstraße 3/1832, A-1040 Vienna, Austria.
e-mail: {bis,ales}@prip.tuwien.ac.at

Summary. In this paper we present a fully automatic and computationally efficient algorithm for optimising multilayer perceptron classifiers. The approach involves two procedures: adaptation (training) and selection. The first procedure adaptively changes the weights of the network. The selection procedure performs the elimination of some of the hidden units (weights). By iteratively combining these two procedures we achieve a controlled way of training and modifying neural networks, which balances accuracy, learning time, and complexity of the resulting network. We demonstrate our method on the problem of multispectral Landsat image classification. We compare our results with a hand designed multi-layer perceptron and a Gaussian maximum likelihood classifier on the same data. Our method produces a better classification accuracy with a smaller number of hidden units than the hand designed network.

1. Introduction

The number of applications of neural networks to remote sensing problems (especially classification) has been constantly rising in the last few years (e.g. see IEEE Transactions Geoscience and Remote Sensing). It has been demonstrated that in many cases neural networks perform considerably better than classical methods e.g. [2]. However, to achieve this, the neural networks need to be carefully designed. This includes both the design of the network topology as well as the input/output representation.

It is now widely accepted, both from a theoretical as well as from a practical point of view, that the degrees of freedom (i.e., the number of weights) influence the performance of a neural network considerably. However, there is no reliable procedure that could a priori give a precise bound on the number of weights or hidden units for a given application. Therefore in most practical situations, one has to adopt a trial-and-error strategy in selecting the number of hidden units (weights). There exists a variety of ad-hoc procedures which determine the network size either by pruning or growing. The algorithm we present here falls into the class of pruning methods. However, we base our algorithm on a firm theoretical basis by considering information theoretic measures to evaluate the complexity of the network. The resulting algorithm

---

* This work was supported by a grant from the Austrian National Fonds zur Förderung der wissenschaftlichen Forschung (No. S7002MAT).

is fully automatic and computationally efficient. The approach involves two procedures: adaptation (training) and selection. The first procedure adaptively changes the weights of the network. The selection procedure performs the elimination of some of the hidden units. By iteratively combining these two procedures we achieve a controlled way of training and modifying neural networks, which balances accuracy, learning time, and complexity of the resulting network.

The structure of the paper is as follows: In the next section we motivate the optimisation principle and describe the objective function of the selection procedure, as well as how we optimise the objective function. Section 2 also includes a description of the complete algorithm. In section 3 we present the experimental results and compare our method to a hand designed network and a Gaussian classifier. Finally we give a conclusion and outlook of further research.

2. Method

In this paper we consider standard 3 layer multilayer perceptrons (MLP). The training set $TS = \{(\mathbf{x}^p, \mathbf{t}^p) = ([x_1^p, \ldots, x_L^p]^T, [t_1^p, \ldots, t_M^p]^T) | 1 \le p \le q\}$ consists of q examples. The network output of the j-th output unit is given by:

$$o_j^p = \sigma(\sum_{i=1}^M w_{ij} r_i^p) = \sigma(\sum_{i=1}^M w_{ij} \sigma(\sum_{k=1}^L w_{ki} x_k^p)), \tag{2.1}$$

where σ is the standard sigmoid function ($\sigma(x) = \frac{1}{1+e^{-x}}$). The network is trained to minimise the usual summed squared error:

$$E = \sum_{p=1}^q \sum_{i=1}^M (o_i^p - t_i^p)^2. \tag{2.2}$$

Let us consider the task of the network as "encoding" the training set. The major building blocks for the "encoder" are the hidden units of the network. Considering a neural network from this point of view we can ask about the shortest possible encoding of the training set by the neural network. It is clear that the length of the encoding depends both on the size of the network and the network error. For example, using a network with an excessive number of hidden units results in no error, however the encoding is large because of the large number of hidden units. On the other hand a too small network causes many errors resulting also in a long encoding. A formalisation of this reasoning leads to the principle of minimum description length (MDL) [11] which is the basis of our selection procedure.

2.1 Selection

We define the task of obtaining a subset of hidden units from a larger set of hidden units as a selection procedure which selects those units that result in a simpler network while preserving the output performance.

Selection is performed by optimising an objective function which can be tied to the principle of Minimum Description Length. Intuitively speaking, if there are more hidden units that "cover" the same space, those units should be selected that describe a larger proportion of the data and which contribute less to the overall error.

In the next two subsections we describe the objective function which encompasses the information about the competing hidden units and the optimisation procedure which selects a set of hidden units, respectively.

2.1.1 Optimisation Function. The objective function which encompasses the information about the competing hidden units has the following form[1]:

$$F(\mathbf{m}) = \mathbf{m}^T \mathbf{C} \mathbf{m} = \mathbf{m}^T \begin{bmatrix} c_{11} & \cdots & c_{1M} \\ \vdots & & \vdots \\ c_{M1} & \cdots & c_{MM} \end{bmatrix} \mathbf{m} \ . \qquad (2.3)$$

Vector $\mathbf{m}^T = [m_1, m_2, \ldots, m_M]$ denotes a set of hidden units, where m_i is a *presence-variable* having the value 1 for the presence and 0 for the absence of the hidden unit i in the resulting network. The diagonal terms of the matrix $\mathbf{C}$ express the cost-benefit value for a particular hidden unit i

$$c_{ii} = \mathrm{K}_1 n_i - \mathrm{K}_2 \xi_i - \mathrm{K}_3 N_i \ , \qquad (2.4)$$

where N_i is the cost of specifying a hidden unit, n_i is the summed activation of the hidden unit $(n_i = \sum_{p=1}^{q} r_i^p)$, and ξ_i is the error caused by that hidden unit $(\xi_i = \sum_{p=1}^{q} |\frac{\partial E}{\partial r_i^p}|)$. The coefficients K_1, K_2, and K_3 adjust the contribution of the three terms. The coefficients can be determined automatically; K_1 is related to the average cost of describing an example (in bits), K_2 is related to the average cost of specifying the error, and K_3 is related to the average cost of specifying a hidden unit. One should note that this term can be used for networks which use different kinds of hidden units e.g., sigmoids, different kinds of radial basis functions. Due to the nature of the problem, i.e. finding the maximum of the objective function, only the relative ratios between the coefficients play a role, e.g. $\mathrm{K}_2/\mathrm{K}_1$ and $\mathrm{K}_3/\mathrm{K}_1$ [9].

The off-diagonal terms handle the interaction between hidden units

$$c_{ij} = \frac{-\mathrm{K}_1 |R_i \cap R_j| + \mathrm{K}_2 \xi_{ij}}{2} \ , \qquad (2.5)$$

$$\xi_{ij} = \max\Big(\sum_{R_i \cap R_j} r_i \xi_i, \ \sum_{R_i \cap R_j} r_j \xi_j \Big) \ . \qquad (2.6)$$

---

[1] Due to lack of space we cannot present a complete derivation of this objective function (see [9]).

$r_{ij} = |R_i \cap R_j| = \sum_{p=1}^{q} r_i^p r_j^p$ is the joint activation of hidden units i and j, and ξ_{ij} is the mutual error of the hidden units defined in Eq. (2.6). In the current implementation we approximated ξ_{ij} by $\max(r_{ij}\xi_i, r_{ij}\xi_j)$.

The objective function takes into account the interaction between different hidden units. However, we consider only the pairwise interactions in the final solution. From the computational point of view, it is important to notice that the matrix $\mathbf{C}$ is symmetric, and depending on the interaction of the hidden units, it can be sparse or banded. All these properties of the matrix $\mathbf{C}$ can be used to reduce the computations needed to calculate the value of $F(\mathbf{m})$.

2.1.2 Solving the Optimisation Problem. We have formulated the problem in such a way that its solution corresponds to the global extremum of the objective function. Maximisation of the objective function $F(\mathbf{m})$ belongs to the class of combinatorial optimisation problems (quadratic Boolean problem). Since the number of possible solutions increases exponentially with the size of the problem, it is usually not tractable to explore them exhaustively. Thus the exact solution has to be sacrificed to obtain a practical one. Various methods have been proposed for finding a "global extreme" of a class of nonlinear objective functions. Among these methods are Winner-takes-all strategy, Simulated annealing, Micro-canonical annealing, Mean field annealing, Hopfield networks, Continuation methods, and Genetic algorithms [3]. We currently use the Winner-takes-all method and a Tabu search algorithm. Tabu search [5] is computationally a little more demanding but it provides consistently better results than the WTA method.

2.2 Complete Algorithm

We can now describe the complete algorithm:

  i. *Initialisation:* We initialise the network with random weights and use a large number of hidden units (e.g. > 20% of the samples).
 ii. *Adaptation:* We adapt the network by a standard training algorithm (in particular, we use a conjugate gradient algorithm). We do not train the network to convergence (see discussion below), usually only a few (e.g., < 10) epochs are sufficient.
iii. *Selection:* Remove redundant hidden units using the selection procedure (section 2.1).
 iv. If the selection procedure does not remove any of the hidden units then train the network to the final convergence, else goto step 2.

This iterative approach is a very controlled way of removing redundant units. Selection is performed based on the relative competition among the hidden units, where we remove only those units that cause higher error and where other units can better generalise the examples. The units which remain in the network are adapted by the training procedure. To achieve a proper selection it is not necessary to train the network to convergence at each step. This is because the selection procedure removes only those hidden units where others can compensate. Since this is independent of the stage of training we can evoke the selection procedure any time[2]. Only after the final selection step is the network trained to convergence. In this way we achieve a computationally efficient procedure.

This is in contrast to other well-known pruning algorithms like Optimal Brain Damage [8] and Optimal Brain Surgeon [6]. They always have to be trained to convergence after one weight/unit is eliminated because only in this case reliable measures of importance for a weight/unit can be obtained. Moreover, these algorithms do not have a stopping criterion, since they only provide a set of networks with decreasing complexity from which the best one has to be selected according to some cross-validation procedure.

One should note that all entries of **C** which we need for the selection step can be calculated during the adaptation procedure, so that this causes no additional costs. Since the matrix **C** is usually very sparse, the selection procedure can be implemented very efficiently, causing only computational costs which are comparable to a single epoch of training.

3. Experimental Results

The data we used for training and testing of the classification accuracy of the neural network were selected from a section (512×512 pixels) of a Landsat TM scene of the surroundings of Vienna. In Figure 3.1a, channel 4 of the test site is shown. This data set has already been used with various other algorithms [2, 1]. Before we describe the results of our method we briefly explain the training data and the hand designed network (for more details see [2]).

The aim of the classification was to distinguish between four categories: built-up land, agricultural land, forest, and water. The resulting thematic map is compared with the output of a Gaussian classification [4]. The Gaussian classifier assumes a normal distribution of the data. Preliminary analysis indicated that this requirement was not fulfilled in the case of the four categories. Therefore, these categories were split into 12 sub-categories with spectral data of approximately normal distributions [12].

Two thematic maps of this scene were prepared by visual classification of the Landsat image, using auxiliary data from maps, aerial photographs, and field work. These two thematic maps showing 4 and 12 land-cover classes were considered to represent the "true" classification of the scene (see Figure 3.1b) and were used for obtaining both training information and test data for assessment of the accuracy of the automatic classification results. The training

---

[2] Except when we have a pathological initialisation (e.g. all hidden units have the same weights, or are confined to a small subspace). In this cases a few steps of training are sufficient to overcome the problem.

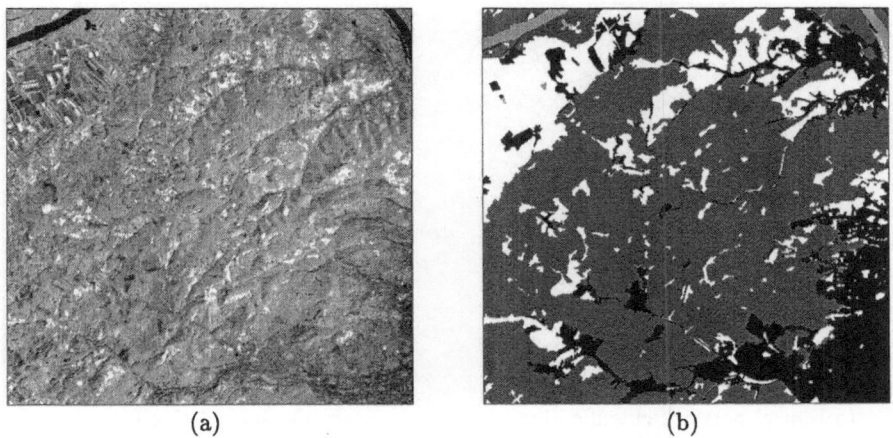

(a)   (b)

Fig. 3.1. (a) Channel 4 of test image and (b) Visual classification

set for both the neural network and the Gaussian classifier consisted of 3,000 pixels.

The hand designed network used for classification is a three–layer feed–forward network with 5 hidden units trained with a conjugate gradient algorithm. The five hidden units have been found by a trial and error method. The inputs to the network are the seven images from Landsat TM channels (bands). As input representation one-dimensional coarse coding [7] is used. The output of the classification task is coded locally.

The network was trained for approx. 80 epochs (complete presentations of the training set) and achieved 98.1% classification accuracy on the training set. Applying the network to the whole image leads to an average classification accuracy of 85.9%, which is slightly better than the Gaussian classifier which achieved 84.7% correctly classified pixels (training and classification with 12 classes, merging into 4 classes after classification). Table 3.1 compares the neural network classification with the Gaussian classifier. Figure 3.2 shows the results of the neural network classification.

Table 3.1. Classification results of Gaussian classifier (ML), neural network (NN), and optimised neural network (Opt. NN)

category	ML	NN	Opt. NN
built-up land	78.2%	87.5%	85.9%
forest	89.7%	89.9%	91.9%
water	84.7%	95.7%	90.4%
agricultural area	74.1%	70.6%	74.2%
average accuracy	84.7%	85.9%	86.2%

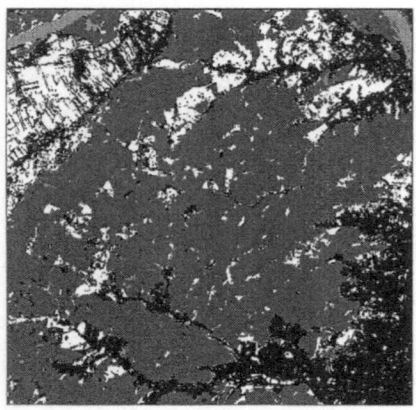
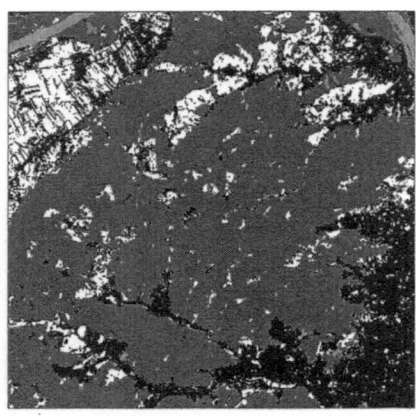

(a Neural network classification)   (b) Optimised classification

Fig. 3.2. Result of neural network and optimised neural network classification (black: built-up land; dark gray: forest; light gray: water; white: agricultural area)

We applied our procedure as described in the previous section to the same data. The starting number of hidden units was 30 (however similar results are obtained with other numbers). The selection procedure was invoked 12 times. The total number of epochs the network was trained is 120. The network ended up with 4 hidden units and the results are shown in Table 3.1 and Fig. 3.2b.

These results show that the fully automatic network setup performs slightly better than the hand designed one with a smaller number of hidden units. The number of epochs is higher than for the hand designed network because due to pruning some retraining has to be performed. However, considering that one has to try more (> 10) alternatives in hand design, we gain also in this case considerable savings in computer time.

4. Conclusions and Work in Progress

We presented a method which starting from an initially high number of hidden units in an MLP network, through an iterative procedure of learning, adaptation of hidden units and selection, achieves a compact network. Our approach effectively and systematically overcomes the problem of selecting the number of hidden units. We have demonstrated that our method performs better than a hand designed network.

The proposed algorithm can be extended in several ways. We have used basically the same algorithm to design radial basis function networks [10] and achieved similar results. We are currently working on adapting the algorithm for several other networks. This includes Gaussian mixture models trained with the EM algorithm and some unsupervised networks. We are also exploiting the possibility to use our algorithm for modular neural network design, and networks which use different types of hidden units. Besides, we plan to develop an incremental variant of the algorithm where we start with a small number of samples and hidden units, and then incrementally add examples and hidden units when needed, and use the selection procedure to remove the redundant ones. Since the method proposed in this paper is quite general, it should be easy to incorporate all these extensions without changing the general paradigm.

References

1. H. Bischof, *Pyramidal Neural Networks*, Lawrence Erlbaum Associates, 1995.
2. H. Bischof, W. Schneider, and A. Pinz, "Multispectral classification of Landsat-images using neural networks", *IEEE Transactions on Geoscience and Remote Sensing*, vol. 30, no. 3, pp. 482–490, 1992.
3. A. Cichocki and R. Unbehauen, *Neural Networks for Optimisation and Signal Processing*, New York: Wiley, 1993.
4. R. O. Duda and P. E. Hart, *Pattern Classification and Scene Analysis*, New York: Wiley, 1973.
5. F. Glover and M. Laguna, "Tabu search", In C. R. Reeves, editor, *Modern heuristic techniques for combinatorial problems*, pp. 70–150, Blackwell Scientific Publications, 1993.
6. B. Hassibi and D. G. Stork, "Second order derivatives for network pruning: Optimal brain surgeo",n In S. J. Hanson et al., editor, *NIPS 5*, pp. 164–172, Morgan Kaufmann, 1993.
7. G. E. Hinton, J. L. McClelland, and D. E. Rumelhart, "Distributed representations", In D. E. Rumelhart and J. L. McClelland, editors, *Parallel Distributed Processing*, vol. 1, pp. 77–109, Cambridge, MA: MIT Press, 1986.
8. Y. L. Le Cun, J. S. Denker, and S. A. Solla, "Optimal brain damage", In D. S. Touretzky, editor, *Advances in Neural Information Processing Systems 2*, pp. 598–605, San Mateo, CA: Morgan Kaufmann, 1988.
9. A. Leonardis, A. Gupta, and R. Bajcsy, "Segmentation of range images as the search for geometric parametric models", *International Journal of Computer Vision*, vol. 14, no. 3,pp. 253–277, 1995.
10. A. Leonardis and H. Bischof, "Complexity optimisation of adaptive RBF networks", *13th International Conference on Pattern Recognition*, vol. IV, IEEE Comp.Soc., 1996 (in press).
11. J. Rissanen, "Universal coding, information, prediction, and estimation", *IEEE Transactions on Information Theory*, vol. 30, pp. 629–636, 1984.
12. R. Wagner, *Verfolgung der Siedlungsentwicklung mit Landsat TM Bilddaten (Settlement development with Landsat TM data)*, PhD thesis, Institute for Surveying and Remote Sensing, University of Natural Resources, Vienna, 1991.

High Speed Image Segmentation Using a Binary Neural Network

Jim Austin

Advanced Computer Architecture Group
Department of Computer Science, University of York, York, YO1 5DD, UK
e-mail: austin@minster.york.ac.uk

Summary. In the very near future large amounts of Remotely Sensed data will become available on a daily basis. Unfortunately, it is not clear if the processing methods are available to deal with this data in a timely fashion. This paper describes research towards an approach which will allow a user to perform a rapid pre-search of large amounts of image data for regions of interest based on texture. The method is based on a novel neural network architecture (ADAM) that is designed primarily for speed of operation by making use of computationally simple pre-processing and only uses Boolean operations in the weights of the network. To facilitate interactive use of the network, it is capable of rapid training. The paper outlines the neural network, its application to RS data in comparison with other methods, and briefly describes a fast hardware implementation of the network.

1. Introduction

The advent of new satellites producing tens of megabytes of image data per day has placed a challenge on the image processing community to derive methods that are capable of delivering timely results, and ways to reduce data to a manageable size. Although there are may techniques for deriving useful information from remotely sensed data (RS), it is not clear whether this is possible at reasonable cost, and in a way that the naïve user is capable of exploiting.

This paper focuses on the every day use of RS data by people such as farmers, Government departments, environmental experts etc. who require results from RS data quickly, easily and at low cost. Typical queries would aim to identify and quantify regions of interest (ROI) within images and to calculate the area of these regions. For example, farmers may require indications of crop yields, planning officials may want to know the amount of urban area usage. This places some very great demands on the designers of such systems, which are far from being met.

It is essential that, given a large resource of image data, a user is able to quickly specify what they require from the data. Because of the nature of image data, specifying the required information by a naive user is most simply done by example, i.e. by the user selecting samples of interest in the image areas using a 'point and click' method. We can imagine that a user might use one or two example images to select what they see as important areas of interest, then submit these to a central server where the images are held. These systems would then, using guidance from the user, search large volumes of RS data, and return the results to the user.

There are a number of methods that could be used within this framework to deliver the results to the user. Within these, neural networks are a strong candidate. They provide a single robust method that can be used for many tasks. They are inherently parallel, permitting the possibility of fast operation. They can be trained using a set of examples provided by the user. They offer the potential to be designed for optimal use automatically. Despite these strengths, their full application is limited by their training speed, which becomes very restrictive with large image sizes, typical of ROI problems. This raises the need for highly specialised and expensive neural hardware to deal with the training problem (see chapter by Day in this book).

The main limitation of neural networks stems from the methods used to train the networks. These rely on techniques that search for an optimal solution to the problem. Because the space of possible solutions increases, at worst, exponentially with the dimensionality of the problem, the training times can become quickly impractical. A number of researchers have realized this and are beginning to propose faster training methods (See Cruse et al. in this book). The work described here has addressed this problem, firstly by the use of a simple training method that scales well with the dimensionality of the problems, and secondly by ensuring that the method can be run very quickly on current digital computers through its use of binary logic operations, or faster on relatively low cost add-on hardware.

The paper first describes the neural network used in the work and then goes on to summarise comparative work that has assessed the method. The final section briefly describes the implementation of the network in high speed hardware and presents some indications of performance.

2. The Binary Neural Network

Reducing the training time of a neural network is essential for their practical use in RS. As outlined in the introduction, conventional network learning is based on an optimisation method that searches the weight space of a neural network to find the set of weights that will minimise some error criterion. Although this can result in a robust classifier, the search time can be excessive. It is well known that other methods of image recognition can get quite good classification results at some cost to recognition time. For example, the K nearest neighbour (K-nn [1]) method is a particularly simple, but relatively successful method of image classification. In simple terms, the k-nn method uses a set, S, of example images for each classification. These images are specified by the user. The method classifies an unknown image as belonging to class S if k of the examples from S are closer to the example than for any other class. Unfortunately, it suffers from particularly slow recognition times,

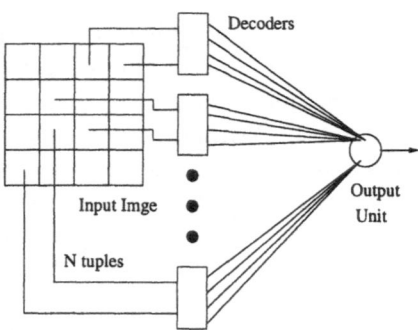

Fig. 2.1. The basic N tuple method.

due to the need to perform (1) a distance measure to all the stored examples, and (2) a sort of the resulting matches to obtain the k closest matching examples. This problem is made worse due to the large numbers of examples needed to get reasonable recognition success. However, setting the method up for recognition is particularly quick.

An alternative method to k-nn, the N tuple method, was developed by Bledsoe and Browning [2]. The N tuple method operates in a similar way to the k-nn method, but does not suffer from long recognition times. It achieves this by combining all the training examples from a class into one template, achieved by the use of a feature pre-processing method called N tuple sampling.

Figure 2.1 shows an outline of the N tuple method. Both training and testing consist of an N tuple pre-process stage. The result of N tuple pre-processing is fed to a storage system based on a neural network.

In the pre-processing stage, the image is broken into a number of samples, called N tuples, each of which is made up of N pixels from the image. Each pixel is taken at random from the image. Each tuple is fed through a function, typically called a decoder, which assigns one of P states to the tuple, i.e. F(N tuple) → state. Each state produced by the decoder function uniquely identifies a combination of pixel values in a given tuple. In essence the N tuple decoder function is a feature recognition system. But, whereas feature recognisers usually identify edges, lines etc. in an image, the decoder function recognises arbitrary pixel value combinations. There are two reasons for this, the first is that it makes no assumptions about the contents of the image, i.e. that edges are the important features, and secondly it allows simple and fast methods to be used to compute the tuple function. Each state feeds directly to an input of a neuron. Note that the state value is typically binary, thus the input to the network is binary. The decoder function used in the current work is described later.

During training, the states produced by the decoder functions are recorded in the neural networks. Each class is assigned an individual single neuron. To train the network a simple hebbian learning method is used. First, the input

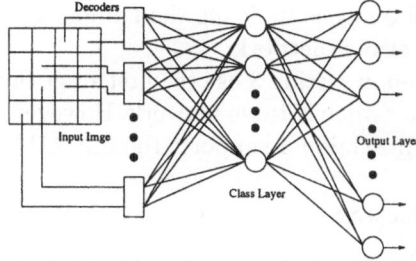

Fig. 2.2. The Advanced Distributed Associative Memory, ADAM.

image is presented and the outputs of the decoder functions computed. These are then fed to the neuron selected for that class of image. Where the input to the neuron is set to one, the weights are set to one (all weights are set to zero initially). Where an input is set to zero, the weight remains unchanged.

The neurons only need record if the tuple state occurred or not. In some applications the frequency of occurrence of the N tuple decoder state over all the training images is recorded. In the simplest case (most often used) no frequency information is stored, only binary 'state has occurred' is recorded. This is to allow very simple and fast implementation in hardware. In many cases the performance loss of this is minimal [3].

Recognition is a simple process of feeding the image sample to the network, passing it trough the decoder functions and then into the network. The usual sum of weights times inputs is used. The result of this passes directly to the output of the network, no activation function is used. The neuron with the highest response indicates the class of the data.

This simple method is fast in training, because all that is recorded is the tuple states, and in recognition only one neuron per class is used making the method particularly fast.

2.1 Extension to the ADAM Memory

The N tuple method was extended to form an associative memory [4], used for image recognition and scene analysis. The aim was to develop a system that could associate an input pattern to another pattern for complex image analysis tasks. For example, the memory could be used to associate image regions with their cartographic equivalents (see section 3.2 for an example of road detection). To allow this, a second layer of neurons was added which allowed the recall of the associated image, and a method of selecting the hidden layer was specified to allow fast training of large numbers of associations.

The architecture of the Advanced Distributed Associative Memory (ADAM) is shown in figure 2.2. The network is trained in 'one shot' as described in Bolt, Austin and Morgan [5] The approach is simple and fast, and allows some degree of analysis to define the number of associations that can be stored for a given size of network.

All the results given in section 3 use the first stage of the ADAM memory, and as such use it as a simple N tuple network. However, the hardware implementation described in section 4 supports the full ADAM implementation.

2.2 Parameter Selection

The introduction indicated that the use of the system should be as simple as possible. In neural network methods that use optimised learning, it is particularly difficult and time consuming to select the parameters that define such things as the number of units, the number of layers etc. This is because no formal model exists for there definition.

The same problem exists with N tuple based systems, in that the user must specify the size of the tuples, the number of tuples to be used and the number of training examples. However, because training is so fast it is possible to define a learning method that cycles through all the possible combinations of parameters to find the optimum. In addition quite a lot is known about the effect of the parameters [6] which can be used to speed up the parameter search.

For the ADAM network the user needs to supply the size of the hidden layer (class or separator layer), as well as the bits set to one in that layer. A great deal is known about the effect of these parameters on the performance of the network [7].

2.3 Strengths and Weaknesses

The networks based on the N tuple method have two great strengths, they can be trained quickly and they can be implemented in conventional computers simply to operate at speed. These advantages come at the cost of recognition robustness. In a MLP or other types of network, the user accepts long training times for possible high accuracy. However, in many applications, such as the one given here, the long training times can result in a system that is inapplicable. It has been recently shown that the N tuple method can result in quite reasonable recognition performance if used with care [8].

It may be noted that there are many extensions to the basic N tuple method, into all types of what is known as Binary, Weightless or RAM based networks [9]. Many of these improve the recognition success at some small cost to training speed and implementation efficiency.

3. Examples of Image Analysis

The N tuple network and the ADAM memory has been used extensively for image analysis tasks [10], [11]. Its application to the analysis of infrared line scan (IRLS) images is of particular interest here [12]. To illustrate

Fig. 3.1. The image used in all examples in this paper. It is a infra-red line scan image taken from 3000ft by an aircraft, 512 × 512 pixels, 8 bits per pixel.

the trade off in accuracy the image in figure 3.1 was used in comparative studies by Ducksbury [13] against a Pearl Bayes Network (PBN) method using conventional image operators. The results of these studies are compared and summarised here.

The images for the comparison are taken from an infra-red scanning sensor taken from an aircraft flying at 3000ft. The area around Luton, UK was used. The data was collected line by line as the aircraft flew along. Each line in the image is one sweep of the IR sensor from horizon to horizon. This results in barrel distortion of the image, and the motion of the aircraft results in line distortions. An example of one of the images is shown in figure 3.1, which is used for the comparisons given in this paper.

3.1 Image Segmentation

Image segmentation is a central task in RS data analysis. The region of interest must be identified by segmenting areas which fall into the same class. This would normally be followed by a area count or some other statistics required by the user.

The problem is often dealt with on a pixel basis, using the multi-band data to classify the pixels into image classes. However, these methods cannot

Fig. 3.2. The results of segmenting the image using the ADAM neural network.

segment regions such as urban areas from rural, due to the contextual information required. The difficult in this task is centred on the amount of data required in terms of the context needed to classify for each region. The most standard approach is to use a small window of pixels to classify a small region of the image. Most approaches use a selection of typical regions with known classifications (ground truth). A convolution (scan) of the whole image selecting regions of the same size for classification is then performed using the system. This is a relatively fast approach compared to other methods which involve diffusion of information from contextual regions to a central pixel.

The simple convolution approach can require an input of up to 256 pixels, or more, to achieve a successful segmentation. For networks such as the MLP the relatively large input window can result in long training times (many hours) on conventional workstations.

The use of ADAM and the N tuple method for segmentation of urban and rural areas in infra-red line scan (IRLS) data is reported in Austin [14]. The method used a gray scale version of the N tuple method first described in Austin (1988) which uses a ranking method in the tuple function. The network was trained on 24 examples of rural and 24 of urban areas, based on 16x16 pixel segments cut from a training image. The training was almost instantaneous on a 10MIPs workstation. The network was then used to classify the rest of the same image, by convolving the network at 16×16 pixel increments. The results are shown in figure 3.2.

Fig. 3.3. The results of segmenting the image using the PBN.

As the network was trained on central regions of the rural area the recognition performance was good in that region, but limited in the right hand region due to distortion of the image.

The same problem was investigated by Ducksbury [13] using a Perl Bayes Network (PBN). The approach first analysed the image using three image filters, creating feature maps for edges, extrema and distribution types. This was then fed to a PBN which was set up to apply a number of relations between the image features. The results of this process are shown in figure 3.3.

The results show that the PBN has developed a clear separation between the urban and field area which is comparable with the neural network based method. Both methods are implementable in parallel hardware for high speed. However, the PBN needed specific features to be selected to allow the correct classification of the regions where as the neural network required no preselection of the relevant features. The advantage of the neural network is clearly in its ability to use raw unprocessed image data. This is highlighted in the next section which shows that the same network can be used to detect road segments with some success.

3.2 Road Detection

Many problems require the annotation of the image to allow identification of the image contents. This is the main task of cartographers. It is a time consuming and expensive process in which all image features must be identified and noted. One such task is road detection, where road like features must be identified and labelled. This is typically a complex task, which first requires the identification of line-like fragments and then the joining together of these into connected line segments. The following summarises the application of ADAM [12] for the initial line segment identification.

The neural network based approach is identical to the segmentation method. A set of example roads were selected by hand and trained into the ADAM network. In this case the ADAM network was set up to recall an icon of the road at the angle given. The data use for training is shown in figure 3.4.

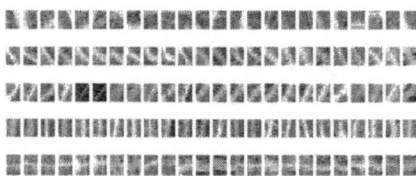

Fig. 3.4. The data used to train the neural network, The data consists of four sets of image patches one for each angle of road (bottom four) and one more non-class data set (top).

The image regions were 8×8 in size, and the gray scale method described in the previous section was used. The result of convolving the image with this network is given in figure 3.5. The method was very good at finding a road if there was one. However, there was a large number of false positives. The method clearly provides a fast pre-search of an image with few false negatives. Subsequent processing on the regions indicated would then provide a robust result.

4. Implementation of the ADAM Network

The N tuple method has been shown to be implementable in dedicated hardware for high speed operation for some time [15]. The essential feature of the method was its similarity to random access memory's (RAM) used in large quantities in computers. The basic binary version of the N tuple method (using a simple 1 in N binary decoder as the tuple function) is basically a RAM. Although it is not practical to use RAMs directly as the neurons and tuple functions, their applicability has given the name to the method (RAM based neural networks).

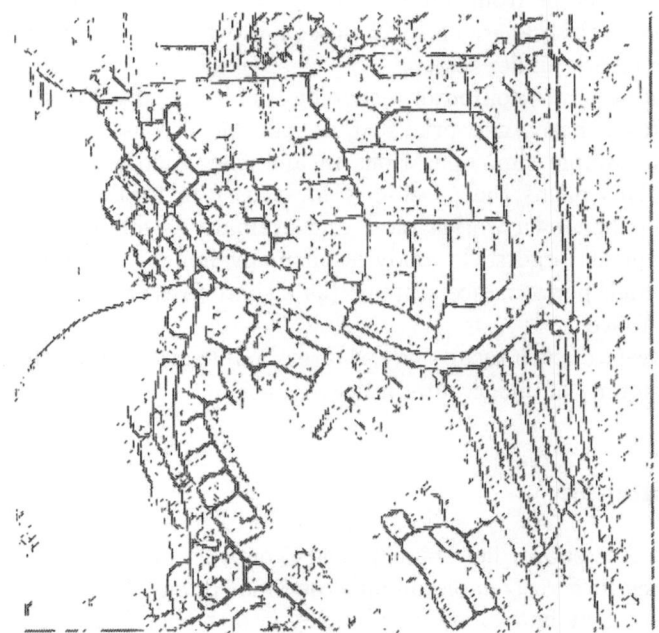

Fig. 3.5. The result of using the ADAM network to find road segments.

Recent work has shown how the weight matrix and recall process can be implemented in field programmable gate arrays [16]. The N tuple pre-processing is undertaken by high performance digital signal processors. The latest implementation of ADAM, using the Sum and Threshold implementation (SAT 2), achieves good performance (using prepared tuple state data). A 512×512 image with an N tuple size 4 can be processed in 32 micro-seconds. The convolution used in the image segmentation task takes in the order of 13 milliseconds. This is equal to the speed of an SGI workstation, using a single MIPS R4600SC, 150MHz processor. The card used to implement the memories is about half the cost of the workstation.

The card used to embed the SAT processor is based on a VME implementation, which allows its use in parallel with a number of other cards, or by itself in a workstation. Our current system is designed to incorporate 3 SAT based cards in a VME based SGI Challenge machine, which contains 4 MIPS R4600SC processors for image handling and pre-processing. The weights are paged from the challenge machine to the cards for processing tasks. In this configuration the machine acts as an image pre-processing engine.

The software to support the ADAM memory is written in C. It allows the user to create, delete, train and test the memories as well as storing and retrieval of trained memories, on a UNIX based work station. A copy of the software is available from our web site [1]. It was used to obtain the results given earlier in this paper. The SAT based hardware uses a C++ based library which is currently under test.

Current standard computer technology is ideally suited to the implementation of RAM based networks such as ADAM. The use of a binary weight matrix and simple sum operations means that good performance can be achieved. This opens the possibility for a user to locally train the neural network on their own workstation, and evaluate its ability on a small test set. Subsequent large scale analysis on many images can then be left for dedicated high performance systems (as a shared resource), such as the one described above.

The advent of such technology at a reasonable price allows the processing of large amounts of image data in near real-time.

5. Conclusion

This paper has shown how binary neural networks can be used to process large images in a reasonable amount of time. It has illustrated that small training sets, may be trained very quickly into the network and achieve very good results. The methods show good performance in comparison with other techniques. The advantage of very high speed implementations of the method provides a direct route to the analysis of large data sets in an interactive environment.

Acknowledgement. This work has been supported in part by the University of York Department of Computer Science; The DTI, EPSRC and British Aerospace under an IED grant; the EPSRC, and British Aerospace within the EPSRC AIKMS initiative. The work has been undertaken in the Advanced Computer Architecture Group by a large number of people, including Guy Smith, Martin Brown, Aaron Turner, Ian Kelly, John Kennedy, Rick Pack and Steven Buckle.

References

1. R. O. Duda and P. E. Hart, *Pattern Classification and Scene Analysis*, John Wiley, 1973.
2. W. W. Bledsoe and I. Browning, "Pattern recognition and reading by machine", *Proceedings Joint Comp. Conference*, pages 255–232, 1959.
3. G. Smith and J. Austin, "Analysing aerial photographs with adam", *International Joint Conference on Neural Networks*, June 7–11, 1992.
4. J. Austin and T. J. Stonham, "An associative memory for use in image recognition and occlusion analysis", *Image and Vision Computing*, vol. 5, no. 4, pp. 251–261, 1987.

---

[1] http://Dcpu1.cs.york.ac.uk:6666/~aaron/adam/adam.html

5. G. Bolt, J. Austin, and G. Morgan, "Uniform tuple storage in adam", *Pattern Recognition letters*, vol. 13, pp. 339–344, 1992.
6. J. Stonham. "Practical pattern recognition", In I. Aleksander, editor, *Advanced Digital Information Systems*, pp. 231–272, Prentice Hall International, 1985.
7. M. Turner and J Austin. "Storage analysis of correlation matrix memories", in preparation, 1996.
8. R. Rohwer and M. Morciniec. "The theoretical and experimental status of the n-tuple classifier", Technical report, Neural Computing Research Group, Department of Computer Science and Applied Mathematics, Aston University, 1995.
9. J. Austin. "A review of ram based neural networks", In *Proceedings of the Fourth International Conference on Microelectronics for Neural Networks and Fuzzy Systems.*, pp. 58–66, Turin, 1994, IEEE Computer Society Press.
10. A. W. Anderson, S. S. Christensen, and T. M. Jorgensen, "An active vision system for robot guidance using a low cost neural network board", In: *Proceedings of the European Robotics and Intelligent Systems Conference, (EURISCON'94) Malaga, Spain*, August 22–25, 1994.
11. D. Bissett, *Weightless Neural Network Workshop '95*. University of Kent, Canterbury, UK., June 1995.
12. J. Austin and S. Buckle, "Segmentation and matching is infra-red airborne images using a binary neural network", in :J. Taylor, editor, *Neural Networks*, pages 95–118, Alfred Waller, 1995.
13. P. G. Ducksbury, "Parallel texture region segmentation using a pearl bayes network", in: John Illingworth, editor, *British Machine Vision Conference*, pp. 187–196. BMVC Press, 1993.
14. J. Austin, "Grey scale n tuple processing", in: Josef Kittler, editor, *Lecture notes in Computer Science, pattern recognition: 4th international conference, Cambridge, U.K.*, vol. 301, pp. 110–120, Berlin, 1988. Springer-Verlag.
15. I. Aleksander, W. V. Thomas, and P. A. Bowden, "Wisard: A radical step forward in image recognition", *Sensor Review*, pp. 120–124, 1984.
16. J. Kennedy, J. Austin, R. Pack, and B. Cass. "C-nnap - a parallel processing architecture for binary neural networks", *ICNN 95*, June 1995.

Efficient Processing and Analysis of Images Using Neural Networks

Stefanos Kollias

Computer Science Division
Department of Electrical and Computer Engineering
National Technical University of Athens
Politechnioupoli, Zographou 15773, Athens, Greece

Summary. A neural network based approach for efficient processing and analysis of images in remote sensing applications is proposed in this paper. Classification, segmentation, coding and interpretation of e.g., Landsat image data are tasks which can take advantage of this approach. First, we use multi-resolution analysis of the images in order to obtain image representations of lower sizes, to which neural networks can be applied more effectively. It is shown that auto-associative neural networks can be used to perform the multi-resolution analysis in an optimal way. Hierarchical neural networks are designed which are able to implement the analysis and classification task at different resolutions. Neural networks are also proposed as an efficient means for selecting regions of interest in the images, further reducing the image representation size which is necessary for effective analysis and classification of the images. Application of the proposed approach to specific real life remote sensing image data is currently under investigation.

1. Introduction

Artificial neural networks have been widely adopted in recent years as a powerful tool for providing intelligent solutions in a wide range of problems and applications. In the fields of pattern recognition and signal and image processing, the nonlinear nature of neural networks, combined with their ability to learn from examples, permit the derivation of effective analysis, classification and diagnosis approaches, even in noisy or time or space varying environments. Various applications, as well as commercial products, based on neural networks, are being developed, such as for recognition of handwritten characters, for analysis of medical images, or for image processing, speech and face recognition and image segmentation.

Analysis and processing of images include tasks such as classification and segmentation, labelling and interpretation, coding and handling of the processed images. The design and use of neural network architectures in these problems is a subject of former and on-going work, which involves many other tasks, such as the image and texture modelling for processing or restoration purposes [1, 3], as well as morphological operators for segmentation tasks [8]. The main focus of this study is on feed-forward neural net works, trained by learning algorithms such as back-propagation, LVQ and probabilistic methods. Recurrent networks, including Hopfield type ones, have also been used in cases, such as deconvolution or restoration problems, which can be formulated as optimisation tasks to which recurrent networks provide efficient solutions [4, 6, 7, 14].

The basic problem when using neural networks for analysis of images is the large size of them, which provides problems of efficiency of training and of the generalisation ability of the networks. The design of efficient versions of learning algorithms [2], as well as the design of structured and modular networks [2, 10, 11] are possible solutions to this problem. These solutions are in accordance with recent theoretical results on feed-forward network generalisation, which refer to pruning or constructive network design, as well as to VC network dimension and the requirement for small sized networks. An important technique for effective processing and analysis of large sized images is the use of multi-resolution/sub-band analysis [15, 16]. The design of hierarchical neural networks which can effectively handle images at various resolutions and combine or interpret the results obtained at various stages of the networks is a very promising neural network based approach to this problem [17, 21]. Another important aspect is the inclusion of invariance in the analysis, classification, labelling, or interpretation of the images. Appropriate neural network architectures can be used which perform these tasks, using either appropriate feature extraction [5, 9, 13] or higher order statistical information [12, 20]. Another important approach to the analysis or coding of such images is the use of appropriately designed neural networks, which can define regions of interest (ROI) in the images and perform the processing, analysis, or coding task mainly focusing on the region of interest parts of the images [18, 19].

In this paper we present a neural network based approach to analysis of images met in remote sensing applications, which results in significant reduction of the size of input images, as well as of the required number of network weights, consequently permitting the effective and efficient classification and interpretation of the images. Section 2 investigates multi-resolution image representations and describes an efficient technique for obtaining low resolution images in an optimal way. Section 3 introduces hierarchical neural networks which are able to analyse and classify images at different resolutions. Section 4 describes the use of regions of interest (ROI) for further reducing the size of the input data, permitting image analysis and classification to be accomplished based solely on the ROI parts of the image.

2. The Multi-resolution Decomposition

2.1 Definition

Representation of signals at many resolution levels has gained much popularity especially with the introduction of the discrete wavelet transform, implemented in a straightforward manner by filter banks using quadrature mirror filters (QMFs) [23]. In image processing the above are equivalent to sub-band processing. Multi-resolution decompositions result in approximation images of low resolution that contain coarse information of the image content and in a set of detail images which contain more detailed information as resolution is gradually increasing.

Let x_0 denote an $N \times N$ image representation. Using appropriate FIR perfect reconstruction filters $h_L(n)$ and $h_H(n)$, where $h_L(n)$ generally is a low-pass and $h_H(n)$ a high-pass filter, we can split the image into four lower resolution images of about $\frac{N}{2} \times \frac{N}{2}$ size [24]. Applying, for example, the low pass filter $h_L(n)$ in the horizontal and then in the vertical direction of the original image we get the approximation image at the lower resolution level $j = -1$, denoted as x_{-1}^{LL}, where

$$x_{-1}^{LL}(m, n) = \sum_{k=1}^{N} \sum_{l=1}^{N} h_L(2m - k) h_L(2n - l) x_0(k, l) \qquad (2.1)$$

By applying all other possible combinations of the above FIR filters, we get three lower resolution detail images, denoted as x_{-1}^{LH}, x_{-1}^{HL}, x_{-1}^{HH}. It is possible to use non-separable analysis (and synthesis) filters to perform the multi-resolution decomposition. In this case, equation 2.1 takes the form

$$x_{-1}^{LL}(m, n) = \sum_{k=1}^{N} \sum_{l=1}^{N} h_{LL}(2m - k, 2n - l) x_0(k, l) \qquad (2.2)$$

Perfect reconstruction of the original image $x_0(k, l)$ can be achieved through synthesis of all four sub-band components. By using only the approximation image $x_{-1}^{LL}(m, n)$ and synthesis filter $f_{LL}(m, n)$ it is possible to obtain an approximate reconstruction $x_0(k, l)$ of the original image as follows

$$x_0(m, n) = \sum_{k=1}^{\frac{N}{2}} \sum_{l=1}^{\frac{N}{2}} f_{LL}(m - 2k, n - 2l) x_{-1}^{LL}(k, l) \qquad (2.3)$$

Optimal design of the analysis and synthesis h and f and filters in specific applications is examined next.

2.2 Neural Network Based Optimal Multi-resolution Analysis

The design of perfect reconstruction filter banks is based on the assumption that all the sub-band signals are available to the interpolation bank with infinite precision. This is not, however, true, when only some of the sub-band components, and particularly only one of them, is used for reconstruction; in this case perfect reconstruction filters lose their optimality. Design techniques for analysis and synthesis filters that perform optimal reconstruction of an original image from a low-resolution representation of it have been recently proposed in [15]. Based on the minimisation of the mean squared error between the original signal and that reconstructed from its low-resolution representation, the $2 - D$ filters are optimally adjusted to the statistics of the input images, so that most of the signal's energy is concentrated in the low resolution sub-band component.

Let us focus next on the problem of generating four sub-band components from each image, only one of which is retained, as the low resolution representation. Let the M-dimensional vector $\mathbf{x}(m, n)$ denote the vectorised $P \times P$ blocks of the input image $x_0(m, n)$, with $M = P^2$, the Q-dimensional vector $\mathbf{y}(m, n)$ denote the corresponding $L \times L$ blocks of the low-resolution representation $x_{-1}^{LL}(m, n)$ also in vectorised form with $Q = L^2$ and finally the M-dimensional vector $\hat{\mathbf{x}}(m, n)$ present the reconstructed vectorised image blocks.

The above vector notations are adopted, so that it is possible to denote the whole convolutional analysis and synthesis operations as multiplications of the above defined vectors by appropriate matrices, say $\mathbf{H}$ and $\mathbf{F}$ respectively. In particular equations 2.2 and 2.3 can be written as

$$\mathbf{y}(m, n) = \mathbf{H}\mathbf{x}(m, n) \tag{2.4}$$

$$\hat{\mathbf{x}}(m, n) = \mathbf{F}\mathbf{y}(m, n) \tag{2.5}$$

Straightforward but tedious calculating, using equations 2.4 and 2.5 provides analytical expressions of the $Q \times M$ $\mathbf{H}$ and $\mathbf{F}$ matrices in terms of the, say $J \times J$, optimal filters h and f respectively. The dimension Q of the low-resolution vector $\mathbf{y}(m, n)$ can be expressed in terms of the input vector dimension M and the length J of filter h. Moreover, the dimension of reconstructed vector $\hat{\mathbf{x}}$ will be greater than the input signal dimension M due to the effects of the synthesis filter length J. In this case, it is the first M elements of $\hat{\mathbf{x}}$ which are retained for comparison with the original vector.

A feed-forward neural network is used next to compute the optimal $J \times J$ analysis and synthesis filters, h and f respectively, through minimisation of the mean squared difference between the original and reconstructed images. The network contains one hidden layer and linear hidden and output units. In particular the network accepts at its input the M-dimensional input image vector x, uses Q hidden units and is trained to produce a reconstructed vector, at its M output units, that is equal to the input vector As a consequence, the network operates in auto-associative form and during training is provided with the same input and desired output image blocks, into which the particular image, or a sequence of images, has been separated; a back-propagation variant with a linear activation function can be the training algorithm. It is desired that the interconnection weights between the hidden units and the network inputs form a matrix W_{IH} equal to matrix H defined above in terms of the optimal filter h, while the interconnection weights between the output and hidden units form a matrix W_{HO} equal to the corresponding matrix F, so that the network implements the operations described in equations 2.4 and 2.5. It should be added that formulations using similar auto-associative linear, as well as non-linear, networks have been proposed for principal component analysis of data. In the following, we impose appropriate constraints in the proposed network architecture, so that it is able to solve the filterbank design problem

Based on the fact [15, 22] that the optimal synthesis filter is related to the analysis one in the frequency domain, or equivalently in the spatial domain

$$f(m, n) = h(-m, -n) \qquad (2.6)$$

the following constraint on the network structure is easily verified

$$W_{HO} = W_{IH}^T \qquad (2.7)$$

Moreover, in order to force matrices W_{IH} and W_{HO} to obtain the required forms, the weights corresponding to zero entries in the matrices are fixed to zero during training. Furthermore when a specific weight of matrix W_{IH} (similarly for W_{HO}) is updated, its value is copied to all other weights that correspond to the same sample value of the optimal analysis filter $h(m, n)$; this procedure is the same as the one used for training time-delay networks, where the need for copying the updated weight values to groups of weights with identical values also arises.

3. Hierarchical Neural Network Classifiers

Hierarchical neural network architectures are proposed next as an efficient scheme for classifying the resulting multi-resolution representations. Let us first assume that a feed-forward multilayer network is used to classify an *approximation* image of quite low resolution. The hierarchical network is then recursively constructed to handle the image at higher resolution levels. More specifically, the proposed procedure starts by training a network at, say, resolution level j with $j \leq -1$, (network LL_j) to classify *approximation* images x_j^{LL} at that resolution level. After training, the network performance is tested, using a validation set of *approximation* images at the same resolution level j. If the performance is not acceptable, training is repeated at the $j + 1$ resolution level.

In this approach, it would be desired that the network at level $j + 1$ (network LL_{j+1}) a-priori includes as much as possible from the "knowledge " of the problem acquired by the former network at level j. Some early results suggested the use of the computed weights of the low resolution network as initial conditions for the weights of the high resolution one. However, we can use the property that the the information of the *approximation* image at level $j + 1$ is equivalent to the information included at both the *approximation* and *detail* images at level j. As a consequence, we can train three more networks (LH_j, HL_j, HH_j), separately (or in parallel) from the former one, to classify the *detail* images at level j and let the network at level $j + 1$ contain in its first hidden layer a number of units equal to the union of the first hidden layer units of all four lower resolution networks.

We then derive forms that permit transfer of the generally large number of (already computed) weights between the input and the first hidden layer of the low resolution networks in corresponding positions of the high resolution network LL_{j+1} and keep these weights fixed during training of the high resolution network. A small number of nodes can be added to the first hidden layer of the LL_{j+1} network, while computation of the resulting new interconnection weights, as well as of the generally less complex upper hidden layers is performed then by training the corresponding parts of the high resolution network LL_{j+1}. To implement the weight transfer, we impose the constraint that the inputs to the units of the first hidden layer of network LL_{j+1} be identical to the corresponding inputs of the units of networks at level j. In the case of network LL_j, for example, with $j = -1$, the input to each unit of the first hidden layer is

$$\sum_{j_1=1}^{L} \sum_{j_2=1}^{L} w_{-1}^{LL}(j_1, j_2) x_{-1}^{LL}(j_1, j_2) \tag{3.1}$$

where $w_{-1}^{LL}(j_1, j_2)$ is the weight connecting each hidden unit to the (j_1, j_2) pixel of the $L \times L$ image block at level -1. Then in the first hidden layer of the LL_0 network classifying each $N \times N$ image block, the input to the corresponding unit will be analogously

$$\sum_{k_1=1}^{N} \sum_{k_2=1}^{N} w_0^{LL}(k_1, k_2) x_0^{LL}(k_1, k_2) \tag{3.2}$$

If the computed values in the above equations are required to be equal to each other, then it can be easily shown that

$$w_0^{LL}(k_1, k_2) = \sum_{j_1=1}^{K} \sum_{j_2=1}^{K} h_{LL}(2j_1 - k_1, 2j_2 - k_2) w_{-1}^{LL}(j_1, j_2) \tag{3.3}$$

A similar form can be derived relating the weights in each network LH_j, HL_j, HH_j and the corresponding weights in network LL_{j+1}. The above forms permit computation of the generally large number of weights between network's input and first hidden layer to be efficiently performed at lower resolution.

It should, however, be mentioned that training and use of all LL, LH, HL, HH networks at level j is not always meaningful; this is due to the fact that in most cases only some of the four sub-band images contain a significant portion of the content of the original image. To overcome this problem, we propose to use only one set of weights, corresponding to the low resolution image which contains the most significant part of the original image among the four sub-band low-resolution images; this image should be created following the procedure presented in the previous section.

4. Region of Interest based Image Analysis

4.1 Definition

All above mentioned strategies, however, rely on metrics and statistics extracted globally over the whole image. In this way, any a-priori knowledge about the image content and spatial variations in the image is disregarded. Schemes that determine regions of interest (ROI) in the images for image coding have been recently proposed in the literature [18]. ROI generally correspond to areas or foreground objects which are of extreme interest for recognition purposes and should be treated with maximum attention; conventional processing or even disregarding of the rest image areas which are of minimal contribution to the perceived information can be adopted in the recognition task. Selection of ROI can be performed, either through user interaction, or automatically when considering specific applications. In the following a neural network architecture is presented for performing the ROI selection.

4.2 Neural Network Based ROI Selection

The ROI selection task is performed by an hierarchical block-oriented two level architecture. In particular, for selecting the ROI regions, the images are first separated in blocks of 8×8 pixels. These blocks are then DCT-transformed, as happens in most DCT-based coding methods. Consequently, classification is performed in the DCT-frequency domain and not in the more-correlated $2 - D$ image space. The first level of the architecture automatically selects all edges appearing in the examined images and classifies the corresponding blocks to the ROI category. This edge selection is due to the fact that, in most applications, the edges existing in the image belong to regions that are of major importance for recognition or classification purposes. This level consists of a feed-forward network which performs the frequency dependent edge detection task, classifying each image block to a 'shade' or 'edge' category. 'Shade' blocks correspond to homogeneous areas containing no significant edges, while 'edge' blocks generally include significant high frequency content. To accomplish this task, the network accepts at its input the computed DCT coefficients of each image block and exploits high frequency information appearing in all components, corresponding, either to different colour or spectral versions of the image.

To let the network detect edges along all different orientations, most, or even all, DCT coefficients from all components of each image block may be required and should be, therefore, presented at the network input, while the number of network outputs is assumed to be equal to two, which correspond to ROI/non-ROI categories.

Supervised learning has been adopted for training the network to perform edge detection. According to it, a predefined training set of characteristic images is selected to which conventional spatial edge detection operators, such as Sobel or gradient ones, or more advanced morphological operators are applied; the results of these operators are further examined by experts to improve the quality of detection. Following this selection the images are divided in blocks which are DCT transformed and labelled as 'shade' or 'edge' ones. The block DCT coefficients for all image components and the corresponding labels for each block are then used to train the network.

After training, the network is able to classify each block of images, that are similar to the ones used for training, to an 'edge' or 'shade' category. In the former case, the block is automatically selected to belong to a region of high visual importance (ROI). If, however, the block is found to belong to an homogeneous region, no decision is taken, but the block is subsequently fed as input to the second level of the proposed architecture, which consists of another network that finally classifies it to a ROI or not. This step is necessary in order to accurately render features like colour or amplitude levels, which, although not corresponding to high frequency information, can be of high importance for classification or recognition.

The second network is similar to the first one, in the sense that it also uses the computed DCT coefficients of the block as input features. The number, however, of these features is generally smaller than the corresponding number of the first network input units, which had to detect edges, i.e., high frequency information across all possible orientations. In particular, the input features that are chosen to feed the second network are the DC coefficient and a small number of AC coefficients following the well-known zig zag DCT scanning of each image component. A supervised learning algorithm has also to be followed in order to train the network, using images from the specific application under consideration. In cases where edges do not play an important role, e.g., in texture-like images, it is possible to overpass the first level of the hierarchical architecture, focusing on the results of the second level.

Whenever a block is classified as one of high importance, i.e. belonging to a ROI, it can be further treated, with maximum accuracy, as the main information/feature to be used for classification or recognition. In the other case, the block may even be disregarded in the following recognition procedure.

5. Conclusions

The above mentioned techniques, and especially multi-resolution image processing and analysis with hierarchical neural networks, region-of-interest selection for coding and analysis of images with neural networks, the design of neural network architectures for efficient classification, segmentation and interpretation of images and signals, can be applied to the processing, coding, analysis and interpretation of e.g. satellite images in remote sensing applications. We are currently aiming at applying these techniques to classification of real life Landsat images.

References

1. Y. Boutalis, S. Kollias and G. Carayannis, "A Fast Multichannel Approach to Adaptive Image Estimation", *IEEE Transactions on ASSP*, vol. 37, pp. 1090–1098, 1989.

2. S. Kollias and D. Anastassiou, "An Adaptive Least Squares Algorithm for the Efficient Training of Artificial Neural Networks", *IEEE Transactions on CAS*, vol. 36, pp. 1092–1101, 1989.

3. S. Boutalis, S. Kollias and G. Carayannis, "A Fast Adaptive Approach to the Restoration of Images Degraded by Noise", *Signal Processing*, vol. 19, pp. 151–162, 1990.

4. S. Kollias, "A Study of Neural Network Applications to Signal Processing", Lecture Notes in Computer Science, vol. 412, pp. 233–243, 1990.

5. S. Kollias, A. Stafylopatis and A. Tirakis, "Performance of Higher Order Neural Networks in Invariant Recognition", in *Neural Networks: Advances and Applications*, pp. 79–108, North Holland, 1991.

6. S. Kollias and D. Anastassiou, "A Unified Neural Network Approach to Digital Image Halftoning", *IEEE Transactions on Signal Processing*, vol. 39, no. 4, pp. 980–984, 1991.

7. S. Kollias and D. Anastassiou, "A Progressive Scheme for Digital Image Halftoning, Coding of Halftones and Reconstruction", *IEEE Journal on Selected Areas in Communications*, vol. 10, no. 5, pp. 944–951, 1992.

8. A. Delopoulos and S. Kollias, "Effective Processing and Analysis of Maps of Iso-Magnetic Curves", Technical Report, NTUA, March 1992.

9. A. Tirakis, D. Kaloyeras and S. Kollias, "A Multimodel Approach to Image Classification Using Neural Networks", *Neural Network World*, vol. 3, pp. 269–287, 1992.

10. D. Kontoravdis, A. Stafylopatis and S. Kollias, "Parallel Implementation of Structured Feed-forward Neural Networks for Image Recognition", *International Journal of Neural Networks*, vol. 2, pp. 91–99, 1992.

11. S. Kollias and A. Stafylopatis, "Parallel Implementations of the Back-propagation Learning Algorithm Based on Network Topology", in *Parallel Algorithms and Architectures for Digital Image Processing, Computer Vision and Neural Networks*, John Wiley and Sons, 1993.

12. A. Delopoulos, A. Tirakis and S. Kollias, "Invariant Image Classification Using Cumulant-based Neural Networks", *IEEE Transactions on Neural Networks*, vol. 5, pp. 392–408, May 1994.

13. L. Sukissian, S. Kollias and Y. Boutalis, "Adaptive Image Classification Using Linear Prediction and Neural Networks", *Signal Processing*, vol. 36 no. 2, pp. 209–232, 1994.

14. D. Foudopoulos, S. Kollias and C. Halkias, "An Efficient Approach to the Detection of Bernoulli-Gaussian Processes for Seismic Signal Deconvolution", *Automatica*, 1994.

15. A. Tirakis, A. Delopoulos and S. Kollias, "2-D Filter Bank Design for Optimal Reconstruction using Limited Sub-band Information", *IEEE Transactions Image Processing*, vol. 4, pp. 1160–1165, August 1995.

16. A. Delopoulos and S. Kollias, "Optimal Filterbanks for Signal Reconstruction from Noisy Sub-band Components", *IEEE Transactions on Signal Processing*, vol. 44, pp. 212–224, February 1996.

17. A. Delopoulos, L. Soukissian and S. Kollias, "Effective Classification of Images at Different Resolution Levels Using Neural Networks", submitted for publication, 1996.

18. N. Panagiotidis, D. Kalogeras, S. Kollias and A. Stafylopatis, "Neural Network Assisted Effective Lossy Compression of Images", *Proceedings of the IEEE*, October 1996.
19. D. Kalogeras and S. Kollias, "Adaptive Neural-network-based Low Bit Rate Coding of Images", submitted for publication, *IEEE Transactions CAS for VT*, 1996.
20. D. Kontoravdis, S. Kollias and A. Stafylopatis, "A Two-Phase Connectionist Approach to Picture Interpretation", *IMACS Transactions, Special Issue on Neural Networks*, to appear, 1996.
21. S. Kollias, " An Hierarchical Multi-resolution Approach to Invariant Image Recognition", *Neurocomputing*, to appear, 1996.
22. P. Baldi and K. Hornik, "Neural Networks and Principal Component Analysis: Learning from Examples Without Local Minima", *Neural Networks*, vol. 2, pp. 53–58, 1989.
23. S. Mallat, "A Theory for Multi-resolution Signal Decomposition: The Wavelet Representation", *IEEE Transactions PAMI*, vol. 79, pp. 278–305, 1991.
24. M. Antonini, M. Barlaud, P. Mathiew and I. Daubechies, "Image Coding using the Wavelet Transform", *IEEE Transactions Image Processing*, vol. 1, pp. 205–220, 1992.

Selection of the Number of Clusters in Remote Sensing Images by Means of Neural Networks

Paolo Gamba[1], Andrea Marazzi[1], and Alessandro Mecocci[2]

[1] Dipartimento di Elettronica, Università di Pavia, Via Ferrata, 1, I-27100 Pavia.
[2] Facoltà di Ingegneria, Università di Siena, Via Roma, 77, I-53100, Siena.

Summary. When processing multidimensional remote sensing data, one of the main problems is the choice for the appropriate number of clusters: despite a great amount of good algorithms for clustering, each of them works properly only when the appropriate number of clusters is selected. As an adaptive version of K-means, the Competitive Learning algorithm (CL) also has a similar crucial problem: different modifications to CL were made with the introduction of frequency sensitive competitive learning (FSCL) and rival penalised competitive learning (RPCL) recently. This last approach introduces an interesting competition mechanism but fails in the presence of real data with multiple clusters of different dimension. We present an improvement of the RPCL algorithm well adapted to work with every kind of real clustering data problems. The basic idea of this new algorithm is to also introduce a competition between the weights in order to allow only one unit to reach the centre of each cluster. The algorithm was tested on multi-band images with different starting weight positions, giving similar results.

1. Introduction

Well known by every researcher involved in the field of data processing is the problem represented by the choice of the number of clusters in multi-dimensional unsupervised data clustering: despite a great amount of good algorithms for clustering, each of them works properly only when the appropriate number of clusters is selected. Even the classical K-means algorithm [1] must be accompanied by some validation algorithms [2] to decide which is the best partition. The problem has been attacked by using statistical approaches, fuzzy approaches and heuristic approaches. Generally all of them have their advantages and disadvantages, depending on the application. More recently neural based approaches have been suggested; for example, as an adaptive version of K-means, the Competitive Learning algorithm (CL) [3] can work as a pre-processing method for clustering analysis and unsupervised pattern recognition. In a competitive learning net, K corresponds directly to the number of neural units used; for good behaviour of the algorithm, this number must be properly selected. Nevertheless CL methods also have some crucial problems, like the presence of dead units [4, 5], when the number of seeds is not properly selected. Various efforts to improve the performance of CL were made with the introduction of Frequency Sensitive Competitive Learning (FSCL) [6] and Rival Penalised Competitive Learning (RPCL) [7]. This last approach completely solves the problem of the choice of the number of clusters but is not robust enough because it works well only in presence of "well behaved" clusters.

In all these kinds of algorithms a certain number of weights is randomly placed in the multidimensional feature space and the system evolves to a situation where some weights are positioned in the cluster centres, while some others are repelled. The use of RPCL applied to artificial data gives very good results and avoids the need of specifying in advance the appropriate number of clusters. Nevertheless, when applied to real data coming from remote sensing images, the RPCL tends to fail. In particular more than one weight falls in the bigger clusters and sometimes, depending upon the initial position, none in the smaller ones.

We present an improvement of the RPCL algorithm called Weight Repulsive Rival Penalised Competitive Learning (WR-RPCL), well adapted to work with every kind of real clustering data problems. The basic idea of this new algorithm is to introduce an additional competition, not only on the clusters point, but also on the weight mutual positions. The algorithm has been tested on multi-band images with different starting positions: for all the tested positions, the algorithm converges to the same steady state.

2. The Evolution of CL Algorithm

In the classical CL algorithm, a certain number of neural units are set in the multidimensional space of the features and iteratively the system evolves, under the effect of forces of attraction, to a situation where each weight is positioned in only one cluster. Given a number of seeds k and denoting ω_i $i = 1, ..., k$ as the randomly initialised weight vector, $\mathbf{x}$ the multidimensional vector representing the data to be partitioned, u_i a coefficient and α_i with $0 \leq \alpha_i \leq 1$ the learning rate decreasing with the number of iterations, the CL algorithm works as follows:

- Step 1: choose a layer of k units, each having output u_i and weight vector, ω_i for $i = 1, ..., k$
- Step 2: randomly take a sample $\mathbf{x}$ from a data set X and for $i = 1, ..., k$ let

$$u_i = \begin{cases} 1 & \text{if } i = c \text{ with } \|\mathbf{x} - \omega_c\|^2 = \min_j \|\mathbf{x} - \omega_j\|^2 \\ 0 & \text{otherwise} \end{cases} \tag{2.1}$$

- Step 3: update the weight vectors by

$$\Delta\omega_i = \alpha_i u_i (\mathbf{x} - \omega_i) \tag{2.2}$$

so that

$$\Delta\omega_i = \begin{cases} \alpha_i (\mathbf{x} - \omega_i) & \text{if } u_i = 1 \\ 0 & \text{otherwise} \end{cases} \tag{2.3}$$

Only the winning unit is updated with the *winner take-all* rule. If the number of seeds is equal to the number of clusters, the result is that each weight moves towards the centre of one cluster. Unfortunately this result is highly affected by the initial position of the weight vectors and very often some weight continues to oscillate between two clusters centres while some other weights remain completely still during the whole iterative procedure, giving rise to the situation of *dead* units.

A good effort for the solution of this problem has been made with the introduction of the FSCL, with a new strategy called *conscience* that takes into account the winning rate of the frequent winner. The CL is therefore modified in the following way:

$$u_i = \begin{cases} 1 & \text{if } i = c \text{ with } \gamma_c \left\| \mathbf{x} - \boldsymbol{\omega}_c \right\|^2 = \min_j \gamma_j \left\| \mathbf{x} - \boldsymbol{\omega}_j \right\|^2 \\ 0 & \text{otherwise} \end{cases} \qquad (2.4)$$

where

$$\gamma_i = \frac{n_j}{\sum_{i=1}^{k} n_i}, \qquad (2.5)$$

n_j is the number of times the weight vector wins upon the considered sample and n_i is the cumulative number of the occurrences of $u_i = 1$.

This strategy is useful to solve the problem of dead units but introduces an other obstacle to a *perfect* clusterisation: when the number of seeds is greater than the number of clusters –situation very common in a real unsupervised case, where the number of cluster is unknown– some weight vectors move towards the centre of the clusters, while some others tend to stop at boundary points between the clusters, greatly affecting the possibility to the use of FSCL as a distance classifier. This algorithms work well only when the number of clusters in the input data set is known in advance and the number of weights is set equal to the number of clusters. This is not so easy, especially when considering multi-band images with no knowledge of a ground map.

A recent modification of FSCL, called rival penalised competitive learning (RPCL) introduces a new mechanism to avoid the problem of the inappropriate number of clusters: the idea is to put in competition the winning weight and the second winner introducing a new pushing force between the two weights, this ensures that each cluster is conquered by only one weight vector. The modification takes place as follows:

$$u_i = \begin{cases} 1 & \text{if } i = c \text{ with } \gamma_c \left\| \mathbf{x} - \boldsymbol{\omega}_c \right\|^2 = \min_j \gamma_j \left\| \mathbf{x} - \boldsymbol{\omega}_j \right\|^2 \\ -1 & \text{if } i = r \text{ with } \gamma_r \left\| \mathbf{x} - \boldsymbol{\omega}_r \right\|^2 = \min_{j \neq c} \gamma_j \left\| \mathbf{x} - \boldsymbol{\omega}_j \right\|^2 \\ 0 & \text{otherwise} \end{cases} \qquad (2.6)$$

and

$$\Delta \boldsymbol{\omega}_i = \begin{cases} \alpha_c \left(\mathbf{x} - \boldsymbol{\omega}_i \right) & \text{if } u_i = 1 \\ -\alpha_r \left(\mathbf{x} - \boldsymbol{\omega}_i \right) & \text{if } u_i = -1 \\ 0 & \text{otherwise} \end{cases} \qquad (2.7)$$

where $0 \leq \alpha_c, \alpha_r \leq 1$ are the learning rates respectively of the winner and the second winner or rival unit. They can depend on time and will always be $\alpha_c(t) > \alpha_r(t)$; γ is the same parameter that was introduced in the previous paragraph.

3. RPCL in Real Problems

The use of the RPCL applied to artificial data gives very good results and avoids the problem of selecting the appropriate cluster number. But in reality it doesn't produce clusters of the same dimension and well separated and the application of the RPCL to this kind of input unfortunately fails. In the case of clusters with a great difference between their dimensions, more than one weight falls in the bigger clusters and sometimes –depending upon the initial position– none in the smaller ones. This fact can be explained by considering that during the competition two weights are attracted from different parts of the same cluster and tend to keep a boundary position with respect to it, in a balance of forces, one of attraction with a part of the cluster and one of repulsion with the competitor weight. An other problem arises when, with clusters of different dimension, the number of weights is equal to the number of clusters: due to the fact that there is a competition, bigger clusters drag two weights, while the smaller ones have not enough strength (number of pixels) to attract any of them. Figure 5.1(a) shows the behaviour of RPCL in a 2-band image, with 2 clusters and 3 initial weights. Figure 5.1(b),(c) shows the behaviour of RPCL with two clusters and three initial weights.

Figure 5.2 shows one of the two-band images 40×40 pixels used in the simulation.

4. The WC-RPCL Algorithm

A first modification is introduced in order to avoid the fact that smaller clusters have not enough strength to attract weights because they can be a second winner with respect to the bigger ones. This happens quite often in multidimensional real images where it is very common to find clusters of different dimension. The basic idea of the modification is to skip the competition between seeds if the second winner is much farther from the cluster than first one. In this way the second winner is not under the interaction of the big clusters and is free to be attracted from other smaller clusters. In this work we consider the competition to be skipped if:

$$\|\mathbf{x} - \boldsymbol{\omega}_c\| > 10 \, \|\mathbf{x} - \boldsymbol{\omega}_r\| \tag{4.1}$$

where $\boldsymbol{\omega}_c$ is the winner and $\boldsymbol{\omega}_r$ is the second winner.

A second modification is introduced in order to avoid the main problem caused by the presence of more than one weight in one cluster. The basic idea of this new algorithm is to introduce a new force to break the balance of forces discussed in the previous section; the competition no longer involves only on the points of the cluster but also the weights. For this, if two seeds are too close, the weakest is repelled farther from the stronger. The evaluation of the strength of each weight is based on the density of pixels surrounding it. So, every weight has a hypersphere of action with a radius depending on the samples of the data set; when an interaction between the hypersphere of each weight occurs, a repulsion mechanism is activated. The first part of the algorithm is the same of RPCL and the modification takes place as follows:

- step 1: find the radius of the hypersphere and the strength of each weight:
  - search the 10 nearest pixels for each weight;
  - find the mean distances between each group of 10 pixel;
  - compute these distances as the radius r of each weight;
  - compute the number of pixels included in the hypersphere of radius r around each weights.
- step 2: if one weight falls in the hypersphere of another weight, it is repulsed with a hyperbolic law:

$$\omega_b = \omega_b + k_i \frac{(\omega_b - \omega_i)}{\|\omega_i - \omega_b\|^2} \qquad (4.2)$$

where k_i is a constant depending on the number of pixels included in the hypersphere.

- step 3: if two competitor weights fall in two hyperspheres respectively:
  - compute the density of pixel for each of the weight;
  - repulse the weakest with the following law:

$$\omega_b = \omega_b + r_i \frac{(\omega_b - \omega_i)}{\|\omega_i - \omega_b\|} \qquad (4.3)$$

where r_i is a constant depending on the number of pixels included in the hypersphere.

The results of this procedure applied to the same 2-band image of figure 5.1 are shown in figure 5.3.

5. Conclusions

The WC-RPCL algorithm allows us to solve the problem of the automatic search for the optimal number of clusters in multidimensional data. The same algorithm was applied to different images obtaining similar results, regardless of the starting position of the weights. The extension of the algorithm to n-dimensional data is straightforward and will be one of the further developments of this work.

References

1. R. O. Duda and P. E. Hart, *Pattern classification and scene analysis*, Wiley Interscience, New York, 1973.
2. X. Zeng, C. Vasseur, "Searching for the Best Partition by Clustering", in *Uncertainty in Intelligent Systems*, Bruchon-Mennier ed., 1990.
3. P. A. Devijver and J. Kittler, *Pattern Recognition: A Statistical Approach*, London U.K: Prentice-Hall, 1992.
4. S. Grossberg, "Competitive Learning: from iterative activation to adaptive resonance", *Cognitive Science*, vol 11, pp. 23–63, 1987.
5. D. E. Rumelhart and D. Zipser, "Feature Discovery by Competitive Learning", *Cognitive Science*, vol. 9, pp. 75–112, 1985.
6. D. Desieno, "Adding a Conscience to Competitive Learning", *Proceedings IEEE International Conference Neural Networks, 1988*, vol. 1 pp. 117–124, 1988.
7. L. Xu, A. Krzyzak, E. Oja, "Rival Penalized Competitive Learning for Clustering Analysis, RBF Net, and Curve Detection", *IEEE Transactions on Neural Networks*, vol. 4, no. 4, pp. 636–649, 1993.

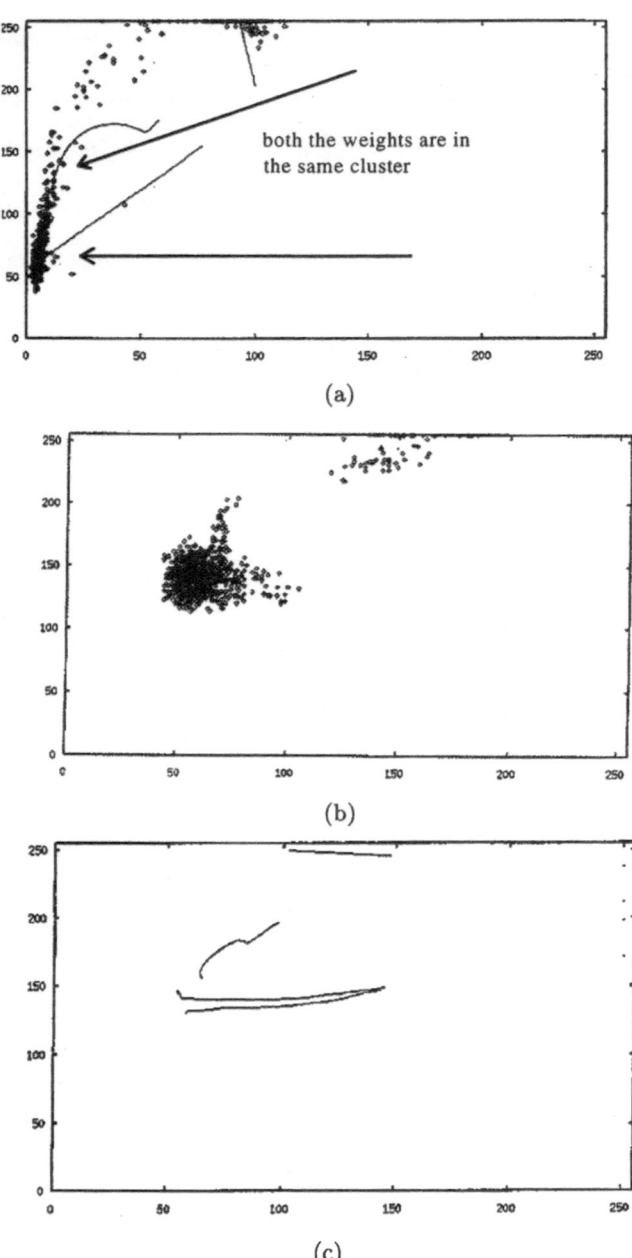

Fig. 5.1. Image with two clusters and three weight: note how one weight is in one cluster, while the others two are positioned in different part of the same cluster; (b) other image with two clusters; (c) three.weights in the big clusters and one the small.

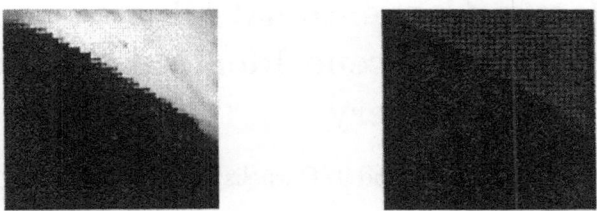

Fig. 5.2. Two band image 40×40 of figure 5.1(a), 8 bit. Two different clusters can be noted

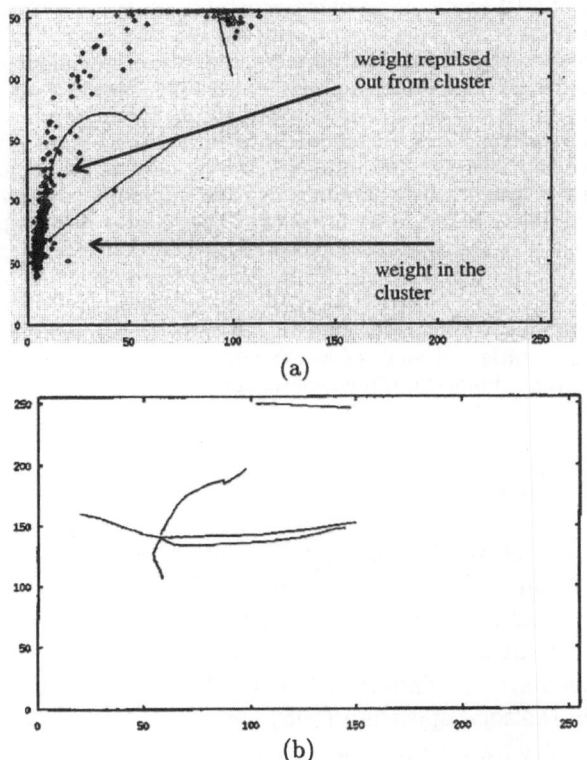

Fig. 5.3. (a) Image with two clusters and 3 weight (see fig. 5.1(a)) It is easy to note how two weights are in two separated clusters while the third weight is repulsed out of the bigger cluster; (b): same result for the problem of figure 5.1(b),(c)

A Comparative Study of Topological Feature Maps Versus Conventional Clustering for (Multi-Spectral) Scene Identification in METEOSAT Imagery

P. Boekaerts[1,2], E. Nyssen[1], and J. Cornelis[1]

[1] VUB - Free University Brussels,
   Department of Electronics, Pleinlaan, 2–1050 Brussels, Belgium.
[2] RMIB - Royal Meteorological Institute of Brussels,
   Department of Aerology, Av. Circulaire, 3–1180 Brussels, Belgium.
   e-mail: pkboekae@etro.vub.ac.be

Summary. A connectionist scheme based on auto-adaptive topological feature maps is compared to conventional cluster analysis for (multi-spectral) scene identification in METEOSAT data. The identification of scenes in (multi-spectral) satellite data is equivalent to the assignment of meaningful labels to spatially coherent image regions. Automated scene detection in METEOSAT data is considered as a data reduction problem with the (optimal) preservation of the spatial coherence of scene region information. A self-organising one-dimensional feature map applied to the so-called segment space of the individual METEOSAT channels is shown to be an appropriate tool for mono-spectral scene identification. It is also shown that the presented connectionist approach for auto-adaptive mono-spectral scene identification has two important advantages compared to conventional cluster analysis, i.e. the number of detected regions is never lower than the number of nodes in the feature map and one obtains a contrast enhancement of the image regions. It is argued however that, despite these properties of feature maps, the use of conventional cluster analysis must be preferred for multi-spectral scene identification because multi-dimensional feature maps become far too slow for practical applications and require a detailed statistical analysis of the multi-spectral segment distribution in order to choose the topological dimension of the feature maps.

1. Introduction

European numerical weather forecasting models rely for an important part on the operational products derived from METEOSAT radiometric image data at EUMETSAT, Darmstadt, Germany. The European geostationary METEOSAT satellites observe the earth with an imaging radiometer in three channels: in the infrared window region (IR) between 10.5 and 12.5 μm (see section 3.1), in the solar spectrum (VIS) between 0.4 and 1.1 μm and in the water vapour (WV) absorption band between 5.7 and 7.1 μm. Images are taken at half hourly intervals and the spatial sampling at the sub-satellite point corresponds to 2.5 km × 2.5 km for the VIS (5000 × 5000 pixels image data, 8 bit/pixel), and 5 km × 5 km in the IR and WV channels (2500 × 2500 pixels image data, 8 bit/pixel).

The automatic MIEC chain image processor (developed at the Meteorological Information Extraction Centre (MIEC), ESOC, Darmstadt) extracts several times a day different types of meteorological information from rectified and calibrated METEOSAT images. The most important meteorological products that cover directly or indirectly the problem of (multi-spectral) scene identification include: Cloud Analysis (CLA), Cloud Top Heights (CTH), Cloud Motion Winds (CMW) and Sea Surface Temperatures (SST).

The cloud classification scheme currently used in the MIEC chain processor is based on multi-spectral histogram analysis within rectangular image regions (windows) of constant size (32 × 32 pixels) [1, 2]. Sub-window cloud classification (the assignment of different cloud classes within one window) is operational (CLA product), but requires manual verification and correction by meteorologists, and can therefore still be improved. Previous studies on NOAA/AVHRR satellite images [3] and research performed by EOS (Earth Observation Sciences) under the direction of the METEOSAT Exploitation Project (ESOC/MEP, Darmstadt) on METEOSAT imagery [4] showed that (neural-like) supervised cloud classification schemes in a feature space deal with the difficulty of optimising the choice of the features, which requires a detailed statistical study of the cloud distributions in the feature space, and the difficulty of collecting a critical amount of relevant training data (manually classified images).The present paper summarises and compares the scene identification results obtained with two non-supervised, auto-adaptive cloud classification schemes, i.e. self-organising topological feature maps for mono-spectral scene identification in the three METEOSAT channels and conventional clustering for tri-spectral scene identification in METEOSAT composite data.

2. (Multi-Spectral) Scene Identification: Methodology

2.1 Scene Identification Task

The identification of a given scene in (multi-spectral) data consists of the localisation of the image region(s) belonging to the scene and the labelling of the(se) region(s). The analysis of scenes on the basis of grey (or colour) values observed on pixel level (pixel analysis) is not sufficient for accurate scene identification because some of the grey (or colour) values are shared by different scene types. If we define a (multi-spectral) scene as a spatially coherent image region with a characteristic (set of) local grey value (or colour) distribution(s), then the task of multi-spectral scene detection consists of assigning one (or a set of unique) value(s) to spatially coherent regions with similar local spectral distribution(s). This is equivalent to assigning one (or a set of unique) value(s) to rectangular sub-regions of the image (or segments) covering a given scene and hence to a data reduction or coding of the (multi-spectral) segment distribution observed in the image [5, 6, 7].

2.2 (Multi-Spectral) Segment Coding

The (optimal) representation (or coding) of the (multi-spectral) segment distribution of a given (composite) image covers the problem of vector quantisation and deals with the question how to find a set of prototypes (code segments) that (optimally) matches a given (unknown) segment distribution. Consider an arbitrary S-dimensional vector distribution V with a given probability density function:

$$P(\mathbf{s}) \mid \mathbf{s} = \mathbf{s}[i,j]$$

The indices i, j are used to treat the corresponding S-dimensional vector of a given (multi-spectral) image segment as a two-dimensional data structure. The problem of encoding may be defined as a mapping of the segment distribution V on a set of S-dimensional code vectors (code segments):

$$V(P(\mathbf{s})) \rightarrow W = [\mathbf{c}_{r1}, \mathbf{c}_{r2}, \ldots, \mathbf{c}_{rn}]$$

This transformation introduces an index mapping function $F(\mathbf{s})$ that maps a given segment on the index of the corresponding code segment:

$$F(\mathbf{s}) = ri$$

with ri being the index of the centroid of a subset of V for which $\mathbf{c}_{ri}$ is coding. An approximation for such mapping can be found by minimising the expected value Ew of the squared error between the segments of the whole segment distribution and the corresponding code segments $\mathbf{c}_{F(\mathbf{s})}$:

$$\left[Ew = \int |\mathbf{s} - \mathbf{c}_{F(\mathbf{s})}|^2 P(\mathbf{s})d\mathbf{s} \right]_{\min} \qquad (2.1)$$

2.3 Topological Feature Maps

Different types of solutions for the corresponding discrete minimisation problem of (2.1) can be found in the literature. Methods based on a local gradient descent technique (applied to a random sampling of the segments from the distribution) include discrete competitive learning (known as the "best match principle" or "*delta*-adaptation"):

$$\Delta\mathbf{c}_{r|\varepsilon \rightarrow 0} = \varepsilon\delta_{r,F(\mathbf{s})}(\mathbf{s} - \mathbf{c}_r) \qquad (2.2)$$

with

$$\delta_{r,F(s)} = 1 \quad \text{for} \quad F(s)=r$$
$$\delta_{r,F(s)} = 0 \quad \text{for} \quad F(s)\neq r$$

An important generalisation of this encoding scheme that allows to link encoding and classification was given by Kohonen [8, 9] who extended the δ-function in (2.2) by a symmetrical, uni-modal function $h_{r,F(s)}$ ("environment function"):

$$\Delta c_{r|\varepsilon\to 0} = \varepsilon h_{r,F(s)}(s - c_r) \tag{2.3}$$

Ritter showed that, if the range of the environment function decreases in time during the adaptation procedure, the repetitive application of the adaptation rule (2.3) to a statistical selection of segments from the segment distribution results in an ordered code segment set for which the following condition holds [10]:

$$\left[Fw = \sum_{r,r'} h_{r,r'} \int_{F(s)=r'} |s - c_{F(s)}|^2 P(s)ds \right]_{min}$$

which may be considered as the generalisation of (2.1). Note that the initialisation of a topological feature map requires the a priori choice of the dimension of the neighbourhood of the code segments (reflected in the environment function). If this dimension (called the dimension of the topological map) is chosen sufficiently large, then the resulting order of the adapted code segments (called a "topological feature map") reflects the statistical properties of the segment distribution and can be used directly as consistent classification labels for the image segments (see section 3.2):

$$s \to ri \to c_{ri}$$

The topological index of a given code segment in a topological map is always a unique label, which implies that the number of detected scenes always equals the number of code segments of the map.

2.4 K-Means Clustering

K-means clustering uses, in contrast to δ-adaptation (2.2), the a priori knowledge of the whole segment distribution (which is the case in most remote sensing applications). The whole code segment set is updated at once after a sequential scan of all the segments belonging to the distribution:

$$c_r = \frac{1}{\sum \delta_{r,F(s)}} \sum \delta_{r,F(s)} \times s \tag{2.4}$$

i.e. the code segments are replaced by the mean segment values of the segments for which the code segments fulfil the best matching principle. K-means reduces for this reason the computational requirements of δ-adaptation. The Euclidean distance between the image segments and the code segments is commonly used as a similarity measure during adaptation:

$$D(\mathbf{s}, \mathbf{c}_r) = \sqrt{\sum_S (\mathbf{s} - \mathbf{c}_r)^2}$$

and iterative repetition of the adaptation procedure (2.4) results in this case in the minimisation of the squared distance error Kw between the whole segment distribution and the code segment set:

$$\left[Kw = \sum_{ij} \sum_k^n D(\mathbf{s}_{ij}, \mathbf{c}_{rk}) \right]_{min}$$

where $\mathbf{s}_{ij}$ indicates a given position of the segment in the (composite) image and where the range of i and j is defined by the image dimensions. As in δ-adaptation (2.2), the code segments are not interrelated after convergence and the code segment indices cannot be used as consistent segment labels (the code segment indices represent random labels in both cases). From equation (2.4), we can easily derive however that after convergence, the central pixel of each code segment represents the mean (multi-spectral) measurement of the central pixels of all the segments for which the code segment is coding. This means that the central pixel of a given code segment can be used, after convergence of the adaptation procedure, as a (consistent) scene classification label representing the mean (multi-spectral) value of the scene for which the code segment is coding. The resulting scene identification procedure guarantees consistent scene labels but does not guarantee that the number of scene labels equals the total amount of code segments used to analyse the data. Different scenes indeed may have the same mean (multi-spectral) value. The reduction of the computational cost for multi-spectral scene analysis by using K-means clustering rather then using topological feature maps results for this reason in a (small) degradation of the scene identification performance.

3. Results

3.1 METEOSAT Radiometric Image Data

The raw METEOSAT images (see section 1) are received and geometrically rectified in real time at EUMETSAT, Darmstadt, before being distributed to the users. The rectified (and inverted) IR channel is the basic input data for the mono-spectral scene analysis presented in this paper and is illustrated in figure 3.1 for slot 27 of July 29, 1992. The IR channel measures thermal radiation in an atmospheric window region emitted from clouds and surface and is inverted to obtain a consistent cloud appearance in the three channels.

3.2 Topological Feature Maps and Mono-Spectral Scene Identification

A digital implementation of data-driven self-organisation of a topological feature map (2.3) to a given vector distribution is described in [5, 6, 8, 9].

The observation that the mono-spectral segment distribution in the IR channel is nearly one-dimensional for horizontal segments of 2×1 pixels [5, 6] is generalised for any (low) segment dimension. This assumption implies that a one-dimensional topological map must be sufficient to represent the METEOSAT mono-spectral segment distributions in the three channels.

The results of auto-adaptive, mono-spectral scene detection in the IR channel of figure 3.1 for a one-dimensional topological map with 36 code segments (N=36) and a segment size of 5×5 pixels are illustrated in figure 3.2. The (one-dimensional) topological indices of the code segments, i.e. the scene identification labels, are represented as rainbow colours (blue= low topological order, red= high topological order). From this experiment (and similar experiment on the VIS and WV channels [5], we can conclude that the code segment indices of a one dimensional topological feature map can be used as consistent scene identification labels (supporting the assumption above). The resulting classified image reveals a contrast enhancement of the original IR channel, which can be understood in terms of segment quantisation and the properties of the topological feature map [5]. A study of the impact of the segment size and the number of code segments on the scene identification results is also presented in [5]. The use of two- and three dimensional feature maps cannot be generalised however for bi- and tri-spectral scene analysis in METEOSAT data. Inconsistent scene identification was observed in both cases. The use of higher dimensional topological feature maps for multi-spectral scene identification becomes impractical however for mainly two reasons: the computational demands of the adaptation procedure of higher dimensional feature maps are far too high for conventional computer hardware and the topological index of the corresponding code segments (the scene identification labels) cannot be represented in the RGB-space anymore.

3.3 K-Means Clustering and Multi-Spectral Scene Identification

METEOSAT RGB composite images of the type VIS (R)-IR (G)-WV (B) represent the basic input data and the evaluation reference for tri-spectral scene analysis of the present study. The resolution of the VIS channel has been reduced by a factor two (2500×2500 pixels) for this purpose, although the full resolution VIS channel can be used in combination with the half resolution IR and WV channels in the multi-spectral scene identification scheme presented in section 2.4. A tri-spectral window of slot 17 of March 7, 1994 is illustrated in figure 3.3. The results of the multi-spectral scene identification scheme of section 2.4 for the tri-spectral METEOSAT data of figure 3.3 is presented in figure 3.4 for 40 code segments (N=40) and a segment size

Fig. 3.1. METEOSAT infrared channel (29-07-1992, slot 27).

Fig. 3.2. Mono-spectral scene detection in the IR channel of figure 3.1 with a 1D topological feature map (N=36).

of 5 × 5 pixels. The classified data in figure 3.4 contains 39 (colour) labels (one less than the number of code segments), which implies that two detected regions have the same mean region value. The consistency of the scene identification results can be verified by comparing figure 3.4 and figure 3.3 (original METEOSAT data). It should be remarked that no contrast enhancement in the classified data is observed any more (compare to the mono-spectral scene identification results in section 3.2). To verify the spatial coherence of the detected scenes, one can compare figure 3.4 with the scene identification results of figure 3.5, obtained after K-means clustering of the RGB distribution (i.e. the tri-spectral pixel data) of figure 3.3. Figure 3.5 shows clearly that the resulting tri-spectral scenes (cloud and surface regions) are fragmented, which is in contradiction with the definition of a scene given in section 1. A study of the impact of the segment size and the number of code segments on the proposed tri-spectral scene identification can be found in [7].

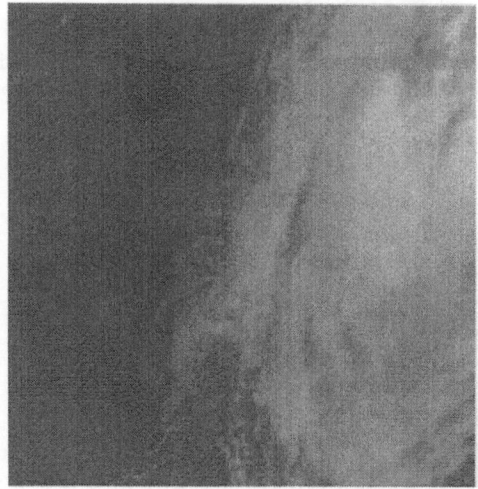

Fig. 3.3. METEOSAT tri-spectral data (window of slot 17, 07-03-1994).

4. Conclusion

The present study has demonstrated that a one-dimensional topological feature map adapted to random segments of individual METEOSAT channels is an appropriate tool for mono-spectral scene identification. The application of two- and three dimensional topological feature maps for bi- and tri-spectral cannot be generalised for bi- and tri- spectral scene identification in METEOSAT composite data. We have shown however that a small extension of

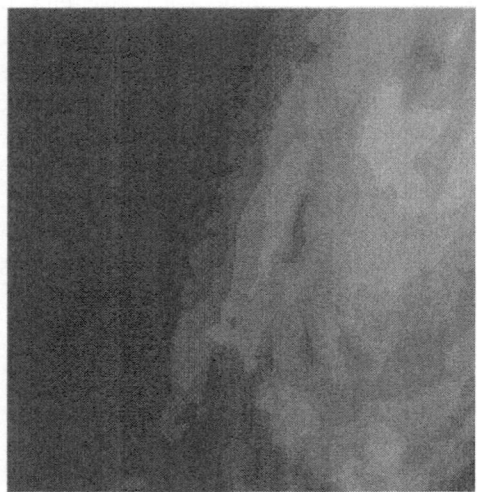

Fig. 3.4. Tri-spectral scene detection in the METEOSAT window of figure 3.3 obtained with conventional clustering of the segment distribution (S=5 × 5, N=40).

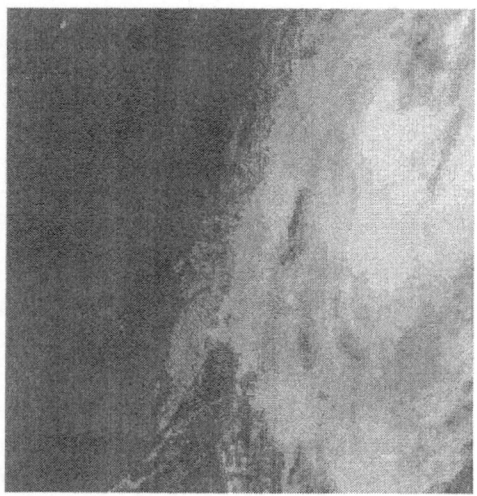

Fig. 3.5. Tri-spectral scene detection in the METEOSAT window of figure 3.3 obtained with conventional clustering in the RGB space (N=40).

the use of a conventional vector quantisation algorithm, i.e. K-means, applied to the multi-spectral segment distribution of METEOSAT composite data, allows to obtain consistent multi-spectral scene identification results with less computational effort and with a minimum loss of classification performance. The proposed multi-spectral scene identification scheme is currently used in a study on the derivation of the angular dependency models for different cloud and Earth surface scenes from polar broad band radiation data by collocation with high resolution tri-spectral METEOSAT data.

Acknowledgement. This research was funded by the Federal Office for Scientific, Technical and Cultural Affairs (DWTC/SSTC) of the Science Policy Office of Belgium, as part of the Telsat III Program (Program Teledetection, contract reference T3/03/41).

The authors thank Mr J. Leber from the European Space Operations Centre in Darmstadt for the METEOSAT data and Mr D. Crommelynck from the Royal Meteorological Institute of Belgium for his support.

References

1. METEOSAT Exploitation Project (MEP), *MIEC Processing*, ESA STR-224, 1987.
2. L. van de Berg, Segment Processing, Internal Report, ESOC/MEP, 1993.
3. P. Boekaerts, E. Steenput, M. Acheroy, P. Van ham and J. Cornelis, "Cloud detection in Multi-Spectral NOAA-AVHRR data using Multi-Layer Perceptrons: a pilot study", *Space Scientific Research in Belgium*, Volume III, Earth Observation, Part 1, pp.173–179, Federal Office for Scientific, Technical and Cultural Affairs, Editor DWTC/SSTC, 1995.
4. EOS, "Neural Networks for METEOSAT Cloud Classification", Final Report, EOS-92/078-RP-004, 1994.
5. P. Boekaerts, Autoadaptive cloud identification in METEOSAT images, *ESA SP-1183*, ISBN 92–9092–331–8, September 1995.
6. P. Boekaerts, E. Nyssen and J. Cornelis, "Autoadaptive mono-spectral cloud identification in METEOSAT satellite images", *EUROPTO, European Symposium on Satellite Remote Sensing*, Conference on Image and Signal Processing for Remote Sensing II, EOS/SPIE vol. 2579, pp. 259–271, November 1995.
7. P. Boekaerts, E. Nyssen and J. Cornelis, "Autoadaptive Scene Identification in Multi-Spectral Satellite Data", accepted for publication for the *5th Symposium on Remote Sensing: a Valuable Source of Information*, NATO AGARD Sensor and Propagation Panel (SPP), 22–26 April 1996, ENSAE-CERT, Toulouse, France.
8. T. Kohonen, *Self-organization and associative memory*, Springer Verlag, Berlin, 1989.
9. H. Ritter and T. Kohonen, "Self-organizing semantic maps", *Biological Cybernetics*, vol. 61, pp. 241–254, 1989.
10. H. Ritter and K. Schulten, "Convergency properties of Kohonen's topology conserving maps: Fluctuations, stability and dimension selection", *Biological Cybernetics*, vol. 60, pp. 59–71, 1989.

Self Organised Maps: the Combined Utilisation of Feature and Novelty Detectors

C. N. Stefanidis and A. P. Cracknell

Department of Applied Physics and Electronics and Manufacturing Engineering, University of Dundee, Dundee, Scotland

Summary. With the interest in neural network implementations for real time applications rising continuously, it becomes important to obtain an insight into their performance and to increase the accuracy of the results obtained by them.

During the classification phase of a neural network a new sample is introduced to the network and the maximum response, i.e. the "best match", is sought. The node with the maximum response then becomes the winner. This method, despite its benefits, provides no indication concerning the degree of match between the new sample and the winning node. Most neural network implementations provide no indication of an input sample that was never "seen" in the past i.e. when nothing similar has ever been used as a sample for training.

To alleviate the above-mentioned problem the Self Organised Feature Detector Map (SOFDM) and the Self Organised Novelty Detector Map (SONDM) were tested. In this paper we present the first results of the combined implementation of the two "conjugate" self organised maps. The SOFDM is a Self Organised Map (SOM) neural network that becomes specifically tuned to the input samples used for training. The SONDM is organised in a way that will result in a small output for "known" patterns, while its output will be very high for "unseen"-novel patterns, thus recognising new patterns.

In an attempt to assess the performance of the networks during their training a measure of disorder is introduced. Based on this measure, different algorithmic approaches are also tested using real data from the National Oceanic and Atmospheric Administration (NOAA) Advanced Very High Resolution Radiometer (AVHRR).

1. Introduction

In the human brain, an almost optimal spatial ordering determined not genetically but based on signal statistics can be established in a simple self organised process under the control of received information. This ordering might be necessary for an effective representation of information. The various maps formed in self organisation describe topological relations of the input signals [1, 2]. Furthermore, some results indicate that some functional principle which operates on a uniform, singly-connected, one-level medium is also able to represent hierarchically related data, by assigning different areas of the storage medium to different abstract levels of information. In most computer simulations the neural network implementation is a one-dimensional or two-dimensional grid; thus the mapping preserves the topological relations while performing a dimensionality reduction of the representation space. The map is even able to represent hierarchical relations between the input data.

2. Formation of Topological Maps of Patterns

The adaptive physical systems used for information processing receive a set of input signals and produce a set of output signals. The transformation of the signal depends on the variable-adjustable internal state. Self organised maps have a particular geometrical configuration which is suitable for display purposes since the units are arranged in a one-dimensional or two-dimensional lattice-array. Each unit-neuron is represented as a point in the lattice, is connected to the external environment and receives the input signal applied. As a result each unit processes the incoming signal and responds with a value in its output. While the input signals change until all members of the sample set are presented to the network, the statistical representation of the data causes the system to change its internal states. These changes are governed by certain principles and rules.

2.1 Construction of Self Organised Feature Maps - SOFMs

A mechanism was proposed in [3] by means of which a neural network can organise itself in such a way that neurons are allocated to a region in the parameter space of external stimuli according to their amount of activity. The aim is to teach the neurons to respond not only to the particular input stimuli with which the network has been trained but to a rather broader range of signals, similar to the inputs the net has experienced during training.

Mathematically, the formation of the topological map can be described as follows. Given an input vector s taken from a space $\mathbf{V}$ of arbitrary dimension l it is mapped to a specific location $\mathbf{m}$ in a two-dimensional space $\mathbf{M}$ which is embodied by a grid of discrete map nodes. The map defines a function $\mathbf{W}$, where:

$$\mathbf{W} : \mathbf{x} \in \Re^l \Rightarrow \mathbf{m} \in \mathbf{M} \tag{2.1}$$

where $\mathbf{m} = (i, j)$ identifies the location of a single node in the map grid. In essence, this algorithm defines a method to find a function $\mathbf{W}$ which preserves information in the mapping from the input space to a grid location [4].

The mapping function $\mathbf{W}$, which is the transfer operator of the neural network, is generated adaptively by adjusting the internal parameters of the map in response to each input vector, thus forming a memory system. The internal parameters consist of a set of weight vectors $w_{(i,j)}$, $i = 1, 2, \ldots, a$ and $j = 1, 2, \ldots, b$ where a, b are the number of nodes of the two-dimensional network. Each weight vector has in total l components, where l is the dimensionality of the input pattern space.

Each node $n_{(i,j)}$ receives the input signal s_t through its connecting weights $w_{(i,j)}$ and forms the discriminant function:

$$n_{(i,j)} = \sum_{d=1}^{l} w_{(i,j,d)} s_{(t,d)} = \overline{w}_{(i,j)}^T \overline{s}_t \tag{2.2}$$

A discriminant mechanism operates by which the maximum of the $n_{(i,j)}$ is singled out:

$$n_k = max\{n_{(i,j)}\} \text{ for } i = 1, 2, \ldots \text{ and } j = 1, 2, \ldots \quad (2.3)$$

SOMs belong to the wider group of self organised competitive learning systems. In these systems the neurons of a layer compete amongst themselves to be activated, with the result that only one output neuron will be activated at any one time. The neuron allowed to be activated is called the winner-takes-all. One way to achieve this winner-takes-all competition is by introducing lateral inhibition i.e. negative feedback paths.

As opposed to other network architectures, SOMs consist of two layers of nodes only. The input layer is comprised of simple-identity nodes that act as the interface of the network with the environment. In the second layer, also called the competitive layer, the neurons are distributed in a one-dimensional or two-dimensional pattern. The latter become, through training, selectively tuned to various input patterns or classes of input patterns. The locations of the neurons so tuned tend to become ordered with respect to each other over the lattice forming a topological map of the input patterns [2, 5]. Spatial locations of the neurons in the lattice correspond to intrinsic features of the input patterns. Neurons transform input signals into a place coded probability distribution that represents the computed values of parameters by sites of maximum activity within the map [6].

Lateral feedback is implemented by the negative feedback paths; the strength of these paths is dependent on the lateral distance from the winning node to the point of application. As can be seen in the exemplar network in figure 2.1, two types of connection can be distinguished: the forward connections from the input layer to the competitive layer and the internal to the network or self-feedback. The input signal is applied to the first layer. The weighted sum of the input signals at each neuron is designed to perform feature detection. The neuron produces a selective response to a particular set of input signals. The feedback connections produce excitatory or inhibitory effects, depending on the distance from the neuron.

Following biological motivation [7], the lateral feedback is usually described by a Mexican hat function shown in figure 2.2.

Three distinct areas of lateral interaction can be distinguished:

  i. a short range lateral excitation area
 ii. a penumbra of inhibitory action
iii. an area of weaker excitation that surrounds the inhibitory penumbra

The principal feature that this network exhibits is the tendency to concentrate its electrical activity into local clusters, named by Kohonen as activity bubbles [8], and the locations of these bubbles are determined by the nature of the input signals.

In figure 2.3 an example of a two-dimensional SOM is shown. The areas of interest are the input nodes comprising the input layer, the connecting

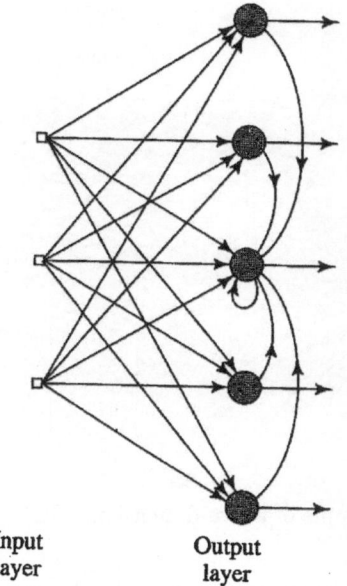

Fig. 2.1. Example of a single layer neural network depicting forward and lateral connections (reproduced from [7])

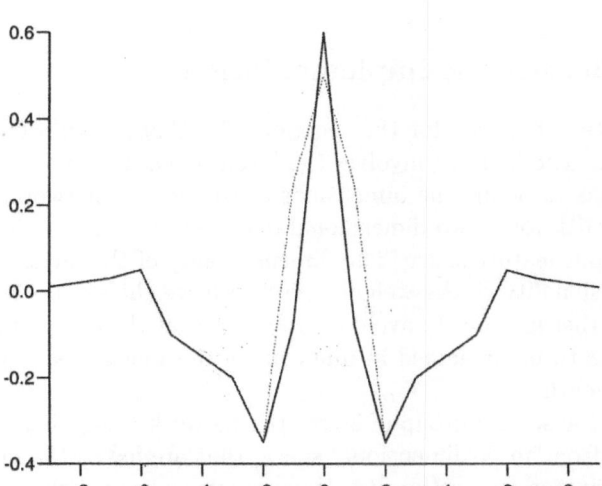

Fig. 2.2. Examples of one-dimensional "Mexican hat" functions

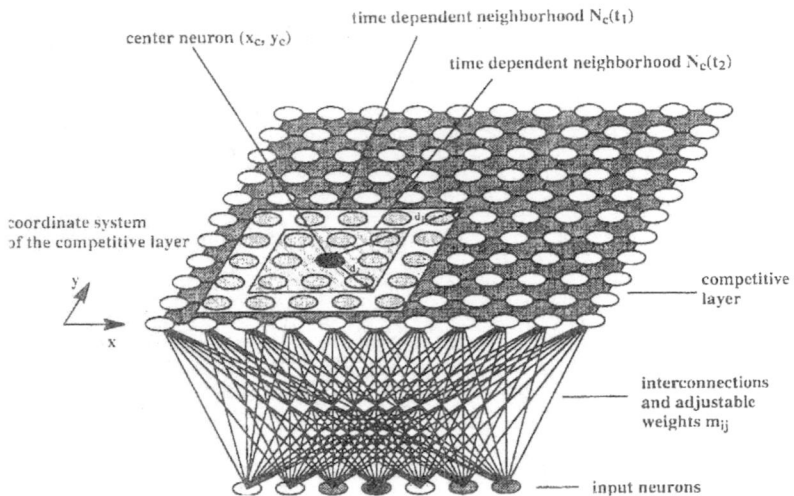

Fig. 2.3. An example of a two-dimensional SOM (reproduced from [9])

weights as straight lines connecting the input layer with the competitive layer, and the competitive layer itself. In the latter the important features are the winning node (shown here as a dark disk), the neighbourhood area (lightly shadowed area surrounding the winning node) and the distance between the winner and the nodes belonging in the topological vicinity of it. The distance is indexed as $d_{1,2}$ identifying distances of one and two nodes respectively.

3. Algorithmic Implementation

The steps followed for the creation of a Feature Map fall into three major groups. The first one involves the formation of the initial weight vectors. The user has to define the dimensions of the desired network, that is the length and width for a two-dimensional net, and also supply the dimensionality of the input feature space. The dimensionality of the latter also determines the dimensionality of the weight vector. All weight vectors must be unequal to each other in order to avoid complications at the latest stage of training. For reasons to be explained in due course, the weight vectors are normalised to unit length.

In the second group of steps, the network is supplied with the samples s taken from an N-dimensional space, that are fed to the input layer, through the weighted connections to the competitive layer. Following a matching rule the weight w_i with the greatest similarity to the input signal is selected. The topological neighbourhood is then defined and the surrounding nodes selected. The update rule is then applied to the winning unit as well as to the surrounding nodes

$$\overline{w}_{(i,j)}(t+1) = \frac{\overline{w}_{(i,j)}(t) + \alpha(t)[\overline{s}_t - \overline{w}_{(i,j)}(t)]}{\|\overline{w}_{(i,j)}(t) + \alpha(t)[\overline{s}_t - \overline{w}_{(i,j)}(t)]\|} \quad \text{for } \forall i \in N_c(t) \quad (3.1)$$

where w_i is the i-th weight vector at iterations (t), and $(t+1)$ respectively, $\alpha(t)$ is the learning parameter with $0 < \alpha(t) < 1$, s_j is the j-th input sample, t is the number of iteration or epoch, and the $\|v\|$ is the Euclidean norm of a vector $\mathbf{v}$.

The prominent feature of this method is the concept of excitatory learning within the neighbourhood around the winning node. The size of this area slowly decreases as the training proceeds.

The final group of steps consists of the retrieving phase, i.e. the stage at which the trained network is used to classify new data samples and therefore utilise the knowledge acquired.

Variations of the updating rule exists and have been tested both in work described in the literature as well as in this project. To incorporate history sensitivity, also helping to alleviate the problem of totally unlearned neurons, the solution is to modulate either the selection of a winner or the learning rate by a frequency sensitivity.

The selection of a winner can be modulated by the frequency sensitivity of the output nodes. As proposed in [10], a history-dependent sensitivity threshold is introduced so that the level of relevant activation is proportional to the amount by which the node exceeds this threshold. This threshold is constantly adjusted. It increases in value whenever a unit loses and decreases otherwise. In this way, if a node is not winning enough inputs the sensitivity increases; thus more nodes are engaged in learning.

4. Novelty Detector

The SOFMs described above belong to the group of feature detectors that become highly activated when an input signal, similar to the ones the neural network has been trained on, is applied to the input layer. A special case of an adaptive unit is governed by the equation:

$$\overline{w}_i(t+1) = \frac{\alpha(t)[-\overline{s}_j - \overline{w}_i(t)]}{\|\alpha(t)[-\overline{s}_j - \overline{w}_i(t)]\|} \quad \text{for } \forall i \in N_c(t) \quad (4.1)$$

where $N_c(t)$ is the topological neighbourhood of the winning node. The most important property of this unit is that if the input patterns are held stationary, the output of the unit will monotonically tend to zero. On the other hand if the input pattern is a new one, then a non-zero output would normally be obtained from the unit. This behaviour is similar to the habilation phenomenon discussed within the context of experimental psychology [2]. This type of SOFM was also reported by Bobrowski [11] where it was called "detector of rareness". The weights as the training proceeds converge to finite, non-zero vectors and the network becomes an orthogonal *projection operator* [2].

Similar behaviour can be achieved for many different input patterns in a set of nodes, the number of which is larger than the number of input patterns. A network with properties of the novel detector was used in this project in conjunction with the SOFMs. In this way, simultaneous training of both networks, i.e. the SOFM and the novelty detector, results in a pair of networks where a high response of the latter indicates that a pattern is unknown and therefore the indication obtained by the SOFM should be ignored.

This tandem approach expands the reliability of the network allowing it to indicate with a certain degree of accuracy the certainty of the response. This assists the network in reducing the number of misclassifications of input patterns.

5. Convergence Analysis, Experiments

In order to obtain an insight on the training process a measure of ordering was needed. Demartines [12] suggested the evaluation of the variance and standard deviation of the distance of neighbourhood nodes to be used as a measure of the level of order of the network. Another measure used in this implementation was the *standard deviation of the winners*.

When a network is specified the weight vectors are set to random initial values. During training these values are then dynamically adjusted to reflect the statistical properties of the input signal. Since topological order is to be preserved, adjacent weights in the grid are expected to be close in value by means of a metric. Based on this, the evaluation of the variances and the standard deviation of the distance of pairs of weights, provides us with an indication of the level of organisation the network has reached.

In simulations using training samples generated by a process with well defined statistical properties this measure was shown to tend to values less than unity.

The standard deviation of the winners was defined as the standard deviation of the difference between the winning node's weight and the input pattern.

A set of simulations was executed using different network specifications. For purposes of comparison the dimensions of the network were fixed to six nodes in length and six in width. The learning rate varied from 0.9 to 0.45 while the network update area varied from covering the complete grid, or just part of it. The set of experiments is listed in table 5.1

The definition of the topological area-Mexican hat function was varied from a wide to a narrow one as can be seen in figure 2.2. The dotted line represents the wide version while the continuous line represents the narrow one. The frequency sensitivity was introduced as a countermeasure to non-learning

Table 5.1. List of the performed experiments.

topological area	frequency	inputs sensitivity	experiments	epochs	learning rate	CPU (sec)
wide	no	pixels (3)	19	5000	0.45	1110[1]
wide	no	pixels (3)	20	5000	0.9	1086[1]
narrow	no	pixels (3)	33	5000	0.9	426[1]
narrow	no	features (12)	42	5000	0.8	1787[1]
narrow	yes	features (12)	43	5000	0.7	474[2]
narrow	yes	features (12)	44	10000	0.7	1048[2]
narrow	yes	features (12)	45	5000	0.8	No record
narrow	yes	features (12)	46	5000	0.8	No record

nodes. This enforces nodes which become winners very often to receive an update inversely proportional to the number of times they won in the past epochs.

In our experiments, we used real data randomly selected from a satellite image. The image originating from NOAA 11, AVHRR was received at approximately 6:00 am on February the 4th, at the Dundee satellite receiving station. Its radiometric resolution was 10 bits while its spatial resolution was 2048 by 2400 pixels. Each pixel has a resolution of approximately 1.1 km at nadir. The image was subsequently sub-sampled creating an image of 512 by 600 pixels (figure 5.1).

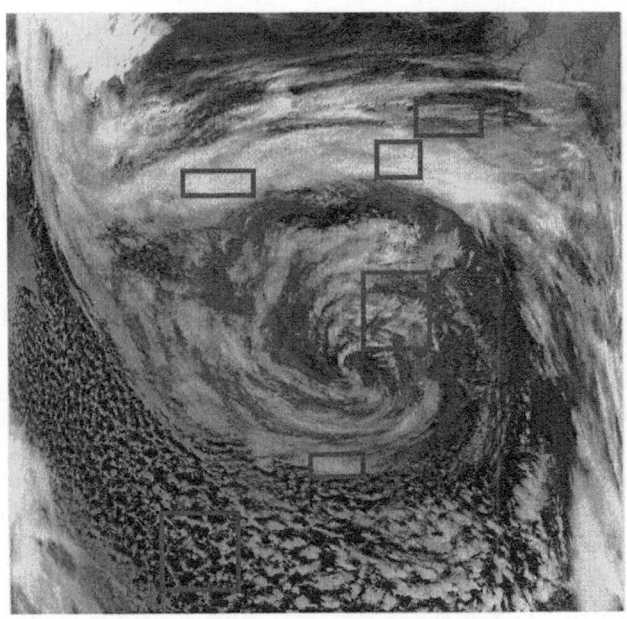

Fig. 5.1. The raw satellite scene with the sampling areas shown as windows

From this image then different inputs were used, shown in the table as *pixels* or as *features*. In the first three experiments depicted on the table the input dimensionality was three, the number in brackets, and only raw-unprocessed data from the three channels of AVHRR scene were used. These included the channel 3, 4 and 5. In the next three experiments the input had a dimensionality of twelve. The last two experiments were a continuation of the training from experiments 43 and 44. For the features with dimensionality of 12 from each pixel position of each channel a moving window was used with a size of five pixels by five pixels and the mean, standard deviation, average entropy and standard deviation of the entropy were extracted. The entropy is a measure of the textural information contained in the window [13]. The fourth column shows the length of training i.e. the number of epochs. The next column shows the value of the learning parameter a used. The last column on this table provides us with an indication of the time required for the network to train. The different subscripts used show two different computer platforms where the experiments were executed. Even when the same platform was used the time length may vary slightly from a perfect linear relation between real time length and number of epochs but this is due to what is known as *overhead load of the system.*

6. Results

Preliminary results indicated instability to occur when while training the network was allowed to reach an update area with one node surrounding the winning node, in both directions. Following the definition of the Mexican Hat function, at this stage the network updates the winning node, as well as the surrounding ones with a positive update rate that varies with the distance. As can be seen in figure 6.1 in the interval between epochs 2000–3000 the disorder level increases and is prevented from reaching a low value. The topological area was reduced to one node surrounding the winning node at epoch 2000 and to zero, i.e. updating only the winner, at 3000 epochs and on.

In order to eliminate this instability the values of the update rate of the nodes surrounding the winning node had to be reduced to a minimum value. In this way the numerical implementation of the *Mexican Hat* function that results, is more narrow in the central area (*see figure 2.2, continuous line*). As a result this instability is only slightly reduced but not eliminated and the network follows a more steady trajectory.

The standard deviation of the distance of the winning nodes from the input patterns seem to provide little if any information on the network progress. In all experiments its value reaches low values soon after the first two thousand epochs while being exponentially reduced with time.

The situation seem to follow a reverse order when the input patterns belong to high dimensionality feature space. As can be seen in experiments 42

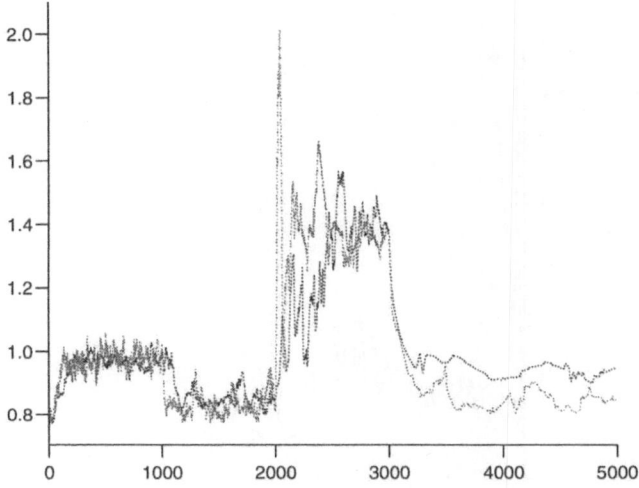

Fig. 6.1. Experiments 19 and 20

and 43 the disorder level increases with time and the network seems to be unable to recognise the same number of classes (figure 6.2). Again the standard deviation of the distance of the winning nodes from the input patterns did not indicate any instabilities or any signs of abnormal behaviour.

In another attempt, the resulting weights from experiments 43 and 44 were used as initial values for a new training session namely experiments 45 and 46 respectively. The network then seem to follow a more steady evolution and eventually reaches low disorder values (figure 6.3). The image segmented using the self organised feature map can be seen in figure 6.4. The areas used to extract the training patterns are also depicted.

7. Conclusions

The implementation of self organised feature maps for satellite images provide us with a fast method to segment the image into disjointed parts with sufficient coherency. The segments created were found to bear also a physical meaning since areas depicting clouds with different cloud tops, i.e. different temperatures, were mapped in different areas of the map. If classification is the aim then in addition to the segmentation algorithm a classification algorithm, either supervised or unsupervised, should be used in conjunction.

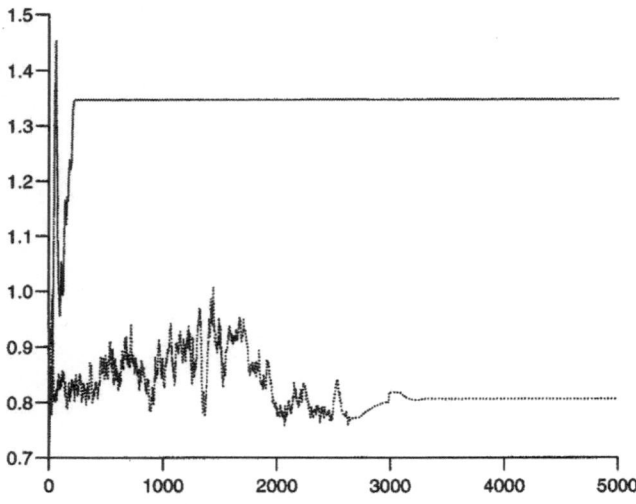

Fig. 6.2. Experiments 42 and 43

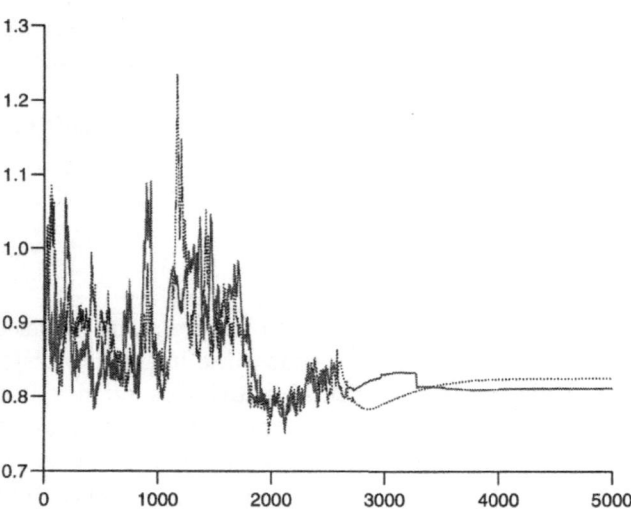

Fig. 6.3. Experiments 45 and 46

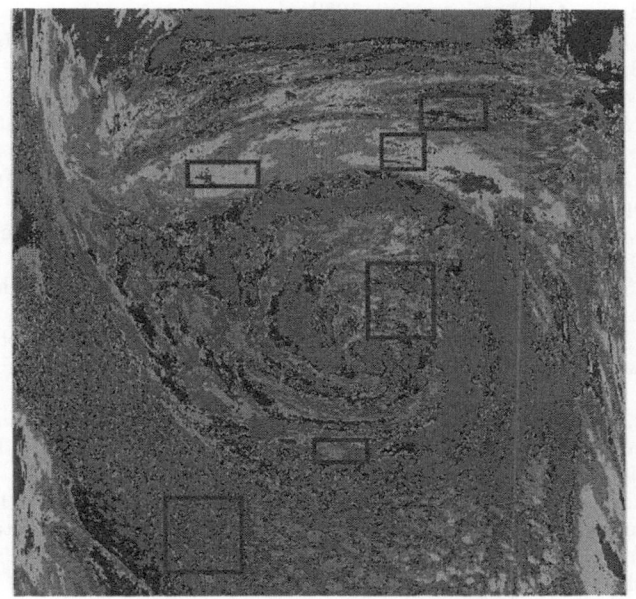

Fig. 6.4. The segmented scene and the colour Look Up Table (LUT) used.

References

1. Kohonen, T., "Self-organised formation of topologically correct feature maps", *Biological Cybernetics*, vol. 43, pp. 59–69, 1982.
2. Kohonen, T., *Self-organization and Associative Memory*, 3rd ed. Springer-Verlag, Berlin, Germany, pp. 312, 1989.
3. MÜeller B., and Reinhardt, J., *Neural Networks an introduction*, Springer-Verlag, Berlin Heidelberg, pp. 266, 1990.
4. Holdaway, R. M., "Enhancing supervised learning algorithms via self organization", in *Proceedings of the International Joint Conference on Neural Networks*, vol. 2, pp. 523–529, 1989.
5. Kohonen, T., "Statistical pattern recognition revisited", *Advanced Neural Computers*, pp. 137-144, 1990.
6. Knudsen, E. I., duLac, S., and Esterly, S. D., "Computational maps in the brain", *AnnualReview of Neuroscience*, vol. 10, pp. 41–65, 1987.
7. Haykin, S., *Neural networks: a comprehensive foundation*, McMillan College Publishing Company, Inc., U.S.A., pp. 696, 1994.
8. Kohonen, T., *Learning vector quantization*, Thesis, Helsinki University of Technology, Laboratory of Computer and Information Science, report TKK-F-A-601, 1988.
9. Gross, M. H., "Subspace methods for the visualization of multidimensional data sets", *Scientific visualization: advances and challenges*, Academic Press Limited, Hong Kong, pp. 171–186, 1994.
10. Bienenstock, E. L., Cooper, L. N., and Munro, P. W., "Theory for the development of neuron selectivity: orientation specificity and binocular interaction in visual cortex", *Journal of Neuroscience*, vol. 2, pp. 32–48, 1982.
11. Bobrowski, L., "Rules of forming receptive fields of formal neurons during unsupervised processes", *Biological Cybernetics*, vol. 43, pp. 23–28, 1982.
12. Demartines, Pierre, Blayo, and Francois, "Kohonen Self-organising maps: is the normalisation necessary", *Complex Systems*, vol. 6, pp. 105–123, 1992.
13. Haralick, R., Shanmugam, K., and Dinstein, I., "Textural features for image classification", *IEEE Transactions on System, Man, and Cybernetics*, SMC-3, vol. 6, pp. 158–165, 1973.

Generalisation of Neural Network Based Segmentation Results for Classification Purposes

Ari Visa and Markus Peura

Helsinki University of Technology
Laboratory of Computer and Information Science
Rakentajanaukio 2 C, FIN-02150, Espoo, Finland

Summary. In this paper the automatic post-processing of segmented images is discussed. The segmentation based on local features and neural networks produces often small regions that disturb further analysis. Strategies for the elimination of these small regions are discussed. One approach based on pyramidal hierarchy is implemented. The approach is tested on a land-based cloud classification problem and the results are reported. This simple strategy applied on the cloud classification problem improves the result 20 - 30 per cent depending on the image. In future continuation of this work it is planned to study how the dynamical expanding context and learning grammars can improve the generalisation of the segmentation and the classification result.

1. Introduction

In computer vision and in remote sensing one of the main problems is the great variety of conditions occurring during the image capturing process. The illumination varies; the size of the object is possibly not large enough to be observed; and the observation may involve a mixture with other objects. This means that the same object may appear to the observer to have different shape or colour depending on the view point, the illumination, and the distance. The problem has been usually controlled by regularisation of the conditions for image capture, by compensating for the distortions with a model, by image enhancement, or by feature selection. All these methods have advantages and disadvantages but the common factor is the significant involvement of human expertise.

As one approach to solve the problem, neural network based image segmentation methods were introduced [7]. However, it was soon noticed that neural networks will not solve the problem by themselves. A combination of pre-selected features and neural networks, especially Self-Organising Feature Maps (SOFM) [5], were found to be an effective tool [8]. The role of the SOFM is to cluster long feature vectors. The result is often good. This means that the segmentation gives large, smooth regions, regions that can be directly interpreted as certain classes. It is also possible that there will be plenty of small regions. These regions are due to the variety of conditions during the image capturing process [9]. For instance in cloud classification -that is our benchmark problem- a cloud might be partitioned into several regions. The central part of the cloud creates typically one large and smooth region and at the edges there are several small regions. All these regions should be merged into one cloud that can be classified.

2. Some Different Strategies

An outline of an automatic image segmentation and classification process is shown in figure 2.1. The location of the actual problem is after segmentation and before the final classification. For the generalisation the use of the shown approach is not required, but the prototypes in the codebook must locally satisfy a certain similarity measure.

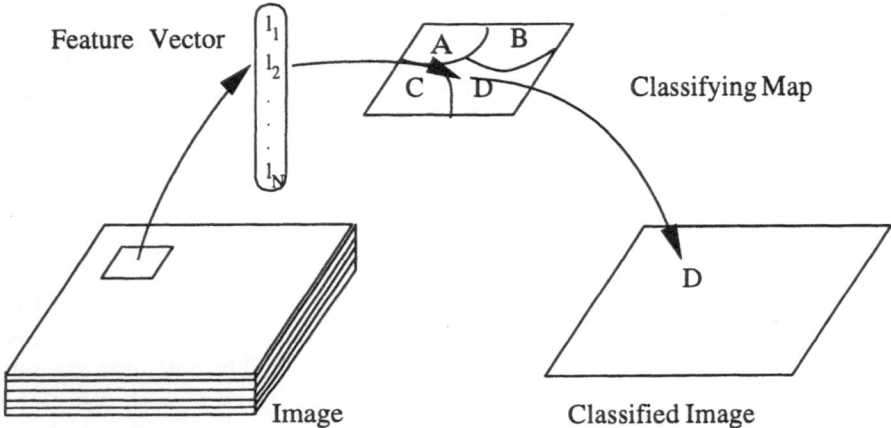

Fig. 2.1. An approach to automatic segmentation and classification. The classification and segmentation are combined

In our case the image is processed with a set of local operators. The image is processed in a scanning manner. The outputs from the operators are concatenated to a feature vector. The feature vector is compared with a set of prototypes, a codebook [7]. The label of the most resembling prototype is assigned to the current location in the output image. If the feature map is not labelled the address of the most resembling prototype is returned. A method for creating compact codebooks by feature maps is presented in reference [5].

After the described pixel-by-pixel segmentation, the resulting image contains pronounced regions of correct labels corresponding to selected properties in the original image. In addition, the image contains disturbances that result from uncertain classifications. Some kind of generalisation is needed to provide a sensible interface upwards. After this generalisation it is meaningful to start the final classification based on shape or size of objects, or hierarchies between objects.

What is then a suitable strategy to create regions? There are two main alternatives. One can start locally, define seed regions, and try to grow the seed regions. This region growing alternative [3] is sensitive to the similarity measure and the selected seed regions. The second alternative is a hierarchic approach. The image is segmented in a pyramidwise way [2]. This is a more successful approach but there are also some limitations. If the object in the image is rotated or translated the pyramid description gives a different data structure. To constrain this problem the image is transferred into polar coordinates. The origin is located on the optical axis. The problem where to start the pyramid structure, is avoided. The image is divided into sectors. The highest level consisting of 16 sectors is shown in figure 3.1. Within a sector majority voting is used. Considering the voting one should remember that similar prototypes support each other.

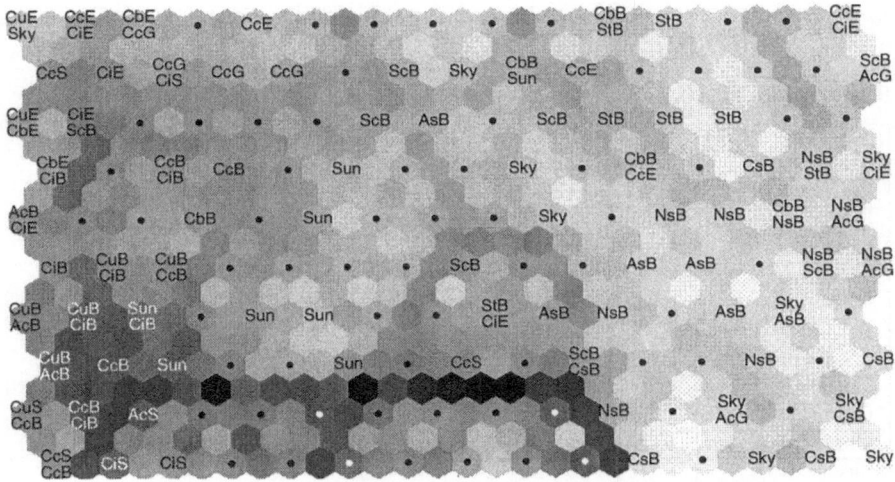

Fig. 2.2. Subclasses projected on the Self-Organising Map

Which segmented image points are similar enough? The problem is solved by using an ordered codebook. The ordered codebook is created by a self-organising process [5]. The adjacent prototypes on the map are similar as indicated in figure 2.2. The similarity is measured by a special distance measure that is described in reference [4].

3. Experiments

The approach has been tested on pictures of clouds. The pictures are captured by a whole-sky near infrared imager and transformed to a digital 512×512 array. The sectors used in generalisation are depicted in figure 3.1. The number of sectors, 16, is chosen so that it is suitable for comparisons with SYNOP observations. The borders of the sectors are based on error analysis to image formation.

Several features are extracted from the source image. Each feature emphasises some distinguishing property of clouds, such as edge sharpness or speck size [6]. As clouds consist of areas with varying appearance, some of the genera are partitioned to subclasses (edge, speck, bulk, gap) that are shown in table 3.1. It should be noted that same subclasses of different genera tend to resemble each other more than subclasses of one genus. The differing subclasses are classified individually in segmentation and finally combined again to construct integral clouds.

Sample images of two cloud genera are shown in figure 3.1. The objective is to distinguish the sun, the sky and the ten cloud genera.

Table 3.1. Partitioning of the target classes.

Target class	Subclasses			
	Edge	Bulk	Speck	Gap
Stratus	StE	StB	'	
Stratocumulus	ScE	ScB		
Cumulus	CuE	CuB	CuS	
Cumulonimbus	CbE	CbB		
Nimbostratus	NbE	NbB		
Altostratus	AsE	AsB		
Altocumulus	AcE	AcB	AcS	AcG
Cirrocumulus	CcE	CcB	CcS	CcG
Cirrus	CiE	CiB	CiS	
Cirrostratus	CsE	CsB		
Sky	Sky			
Sun	Sun			

The pre-classified samples needed in the classification consist of single feature vectors. The sampling can be automatized to a large extent. The collection of samples, a codebook, is labelled by investigating a comprehensive set of cloud images and extracting feature vectors at positions containing representative details of cloud genera. Each feature vector gives a label to a prototype in the codebook.

Some of the classified images are shown in the left of figure 3.1. Out of the 176 classified sectors, 115 (65%) matched the manually given labels. Variation in classification of different cloud genera is shown in the confusion matrix (table 3.2). The generalisation improves the sectorwise classification by 10–15% depending on the image.

Table 3.2. Confusion matrix of the experimental classification.

	True Class											
---	Sky	Sun	St	Sc	Cu	Cb	Ns	As	Ac	Cc	Ci	Cs
Sky	62	0	0	0	0	0	0	0	31	0	22	8
Sun	0	90	0	6	17	0	13	0	0	0	0	0
St	3	0	100	0	0	23	0	0	0	0	0	0
Sc	3	0	0	38	0	23	6	0	0	0	0	8
Cu	3	0	0	13	83	8	0	0	19	0	0	0
Cb	0	0	0	0	0	0	0	0	0	11	0	0
Ns	3	0	0	38	0	15	81	13	0	11	0	8
As	0	0	0	6	0	0	0	88	0	0	0	0
Ac	5	0	0	0	0	0	0	0	44	0	11	0
Cc	3	10	0	0	0	31	0	0	6	78	0	0
Ci	0	0	0	0	0	0	0	0	0	0	67	0
Cs	19	0	0	0	0	0	0	0	0	0	0	75
#	37	10	16	16	6	13	16	16	16	9	9	12

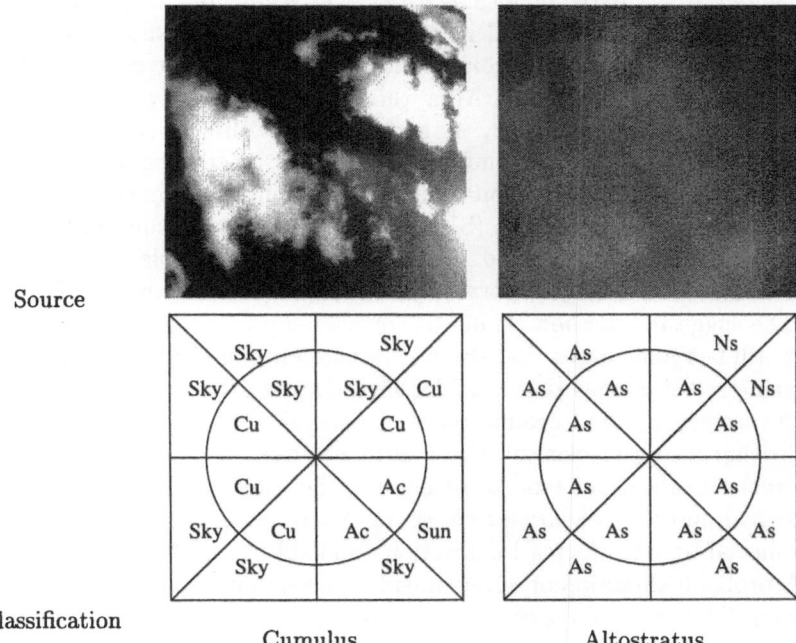

Source

Classification

Cumulus   Altostratus

Fig. 3.1. Some raw all-sky images and the corresponding classified generalisations

The number of sectors labelled to each class is in the bottom row. The percentage distribution of the automated classification is listed by columns. In fact, the main diagonal is rather consistent. The most conspicuous class is certainly cumulonimbus (Cb) without correct classifications. It is expected that cumulonimbus will remain problematic in land-based recognition as its size exceeds the scope of whole-sky imagers. Consequently, the concerned segmented image in figure 3.1 contains acceptable misclassifications. The fragmented edge of the developing cloud does resemble Altocumulus (Ac). The precipitating part of the cloud, approaching from the left, is classified as Nimbostratus (Ns).

4. Discussion

Neural networks have been shown to give useful image segmentations. However, the segmentation result suffers from the same problems as classical clustering approaches [3]. Some post-processing is still needed but the processing should be automatic in the same sense that neural networks are.

The problem has been studied and the results have been demonstrated on the problem of land-based cloud classification. The suggested method treated whole-sky images as a composition of objects. The images were first segmented. It was based on local features. The segmented image were divided into sectors that were treated in pyramid manner. The target classes were also divided into subclasses corresponding to the different textures appearing within each cloud genus. After the initial pixel-wise classification, the subclasses were recombined by majority voting within a sector. The voting results were treated in a pyramid manner to construct integral clouds.

The applied sectored output differs from the international encoding standard of synoptic observations (SYNOP). However, the results of the experimental classification are good. The results are comparable to some extent with those of the classifier reported by Buch, Jr. and Sun in the ARM program [1]. The suggested method should be considered promising in this respect. It must still be remembered that the samples used in this study were taken from the same set of images that was classified. Despite the limited set of source images, some objective evaluation of the classifier can be performed by investigating incorrect classifications as in the case of cumulonimbus. Many incorrect classifications tend to be close to the true classes in both the morphological and the meteorological sense. Many incorrect classifications can be removed by introducing information on cloud altitude.

A probable direction of development is towards layered classification. Elementary classes would serve as an intermediate level of decision. These decisions are combined within a region in statistical sense. The regions are arranged in hierarchies, where pyramid and tree structures are useful. Solutions to some subproblems on the way to the described goal are given in this paper.

According to the experience obtained in this study, clouds and other complex objects, seem to be interpretable by means of the described methodology. This, together with the various potential applications of automated cloud classification, encourages us to continue this research.

References

1. K. A. Buch Jr. and C.-H. Sun., "Cloud classification using whole-sky imager data", *9th Symposium on Meteorological Observations and Instrumentation*, pp 353–358, Charlotte, North Carolina, March 1995.
2. P. Y. Burt, T. H. Hong, and A. Rosenfeld, "Segmentation and estimation of image region properties through cooperative hierarchical computation", *IEEE Transactions on Systems, Man and Cybernetics*, vol. 11, no. 12, pp 80–809, 1981.
3. R. M. Haralick, and L. G. Shapiro, "Image Segmentation Techniques", *Computer Vision, Graphics, and Image Processing*, vol. 29, pp. 100–132, 1985.
4. J. Iivarinen, J. Rauhamaa, and A. Visa, "Unsupervised Segmentation of Surface Defects", to appear in *Proceedings of 13th IAPR International Conference on Pattern Recognition*, Vienna, Austria, August 26–29, 1996.
5. T. Kohonen, "The Self-Organizing Map", *Proceedings of the IEEE*, vol. 78, no. 9, pp. 1464–1480, 1990.
6. M. Peura, and A. Visa, "A New Approach to Land-Based Cloud Classification", to appear in *Proceedings of 13th IAPR International Conference on Pattern Recognition*, Vienna, Austria, August 26–29, 1996.
7. A. Visa, K. Valkealahti, and O. Simula, "Cloud Detection Based on Texture Segmentation by Neural Network Methods", *International Joint Conference on Neural Networks (IJCNN)*, 18–21 November, Singapore, pp. 1001–1006, 1991.
8. A. Visa, "Unsupervised Image Segmentation Based on a Self-Organizing Feature Map and a Texture Measure", in: *Proceedings of 11th IAPR International Conference on Pattern Recognition*, The Hague, The Netherlands, August 30–September 3, pp. 101–104, 1992.
9. A. Visa, K. Valkealahti, J. Iivarinen, and O. Simula, "Experiences from Operational Cloud Classifier Based on Self-organizing Map", in: *Proceedings of SPIE's International Symposium on Optical Engineering in Aerospace Sensing, Vol 2243, Applications of Artificial Neural Networks V*, Orlando, Florida, April 5–8, pp. 484–495, 1994.

Remote Sensing Applications Which may be Addressed by Neural Networks Using Parallel Processing Technology

Charles Day

Environmental Mapping and Modelling Unit, Space Applications Institute, Joint Research Centre, European Commission, 21020, Ispra, Varese,Italy

Summary. More instruments per payload, instruments which are increasingly sophisticated, and a fleet of Earth observation satellites will all conspire to flood ground stations throughout the world with an unprecedented quantity of data by the turn of the century.

This deployment of remote sensing hardware is the response of Earth observation agencies world-wide to the challenge of finding out more about the Earth's natural systems and the stresses placed upon them by human activity.

Acquiring the data however only addresses one part of the challenge. Another equally critical requirement is the need to transform the acquired data into timely, high quality, and cost-effective information that can be used by Government agencies, scientists, and industry to enable better stewardship of the Earth and its resources.

Not only is the quantity and rate of data capture increasing but the trend for performing multi-temporal analyses of, already formidable, datasets is also likely to tax the data-processing facilities of even the best equipped ground stations.

This paper examines how the coupling of High Performance Computer Networks (HPCNs) and the data processing potential of Neural Networks (NNs) may have important consequences for the efficient generation of some Earth observation products and information.

1. Introduction

The latter half of this decade will see an unprecedented increase in the quantity (in excess of 10^{13} bytes weekly) and the quality (more spectral channels, some with a spatial resolution of 2.5 metres) of remotely sensed data.

The management of this data, ie. storage, access and archiving, is a major undertaking which is being addressed by the US *Earth Observing System Data Information System* (EOSDIS) and the *European Earth Observation System* (EEOS). The EEOS incorporates the European Commission's *Centre for Earth Observation* (CEO) project which is managed by the European Commission's Joint Research Centre (JRC) and Directorate General (DG) XII. A primary objective of the CEO project is to allow easy access to remotely sensed datasets and thereby promote their wider use.

The technological challenges presented by the management of such huge quantities of data will necessarily involve the deployment of advanced data handling techniques.

Turning the data into useful information is also a formidable undertaking. Existing algorithms for the manipulation of remotely sensed data to provide useful high quality information about the land, the oceans and the atmosphere are inherently computationally intensive.

Hardware and software developments of recent decades have delivered vastly more powerful computational machinery, almost annually. Even so, practitioners in remote sensing are often confronted with data processing tasks which entail long delays of hours or even days of processing time.

Failure to adequately address the challenges of turning ever larger quantities of remotely sensed data into high quality environmental information products is likely to reduce their attractiveness to customers for whom quality and timeliness is often a scientific or commercial imperative.

One possible means for delivering the necessary improvements in processing speed lies in the use of parallel processing techniques. Such methods require the use of specialised hardware and software which enables separate parts of a large problem to be tackled simultaneously, with the result that the overall processing time is reduced. The provision of parallel processing hardware is no longer a technological problem but fundamental problems exist when large scale parallelism is introduced in complex software systems. However, since the late 1980s an inherently parallel paradigm for the complex transformation of numerical data has been the subject of much computer science research. Neural Network (NN) techniques, inspired by the ease with which mammalian brains seem to be able to process vast amounts of information with relatively low hardware speed (eg. biological neurons fire at a rate of a few milli-seconds compared to the micro-second switching speeds of todays CPUs [1]), may provide a means for addressing this problem.

NN algorithms deliver networks which process data in an inherently parallel fashion. A large problem remains however. Bespoke NNs are usually needed for each data transformation and the time taken for NN training software to deliver each network can be large; NNs are trained iteratively, often using large amounts of training data, rather than being coded directly.

NN training times may be the major obstacle to be overcome if NNs are to become widely used for the data processing needs of the remote sensing community. However, tackling this problem provides a synergistic opportunity for progressing with the European Commission's *Fourth Framework Programme*. This programme includes investigations into the use of High Performance Computing and Networking (HPCN) as an important activity under its *Information Technologies* (IT) programme.

Investigating the extent to which a European HPCN can contribute to the usefulness of NN techniques for the data processing requirements of remote sensing's scientific and commercial user-communities is the subject of this paper.

2. Technological Developments in the Acquisition of Remotely Sensed Data

Two key developments are likely to have an enormous impact upon the quantity and quality of remotely sensed data as the millennium approaches. They relate to the number of new satellites that will be launched in the short to medium term and the sophistication of the sensors carried by each of these platforms.

Future European Space Agency (ESA) remote sensing missions will include:

- Aristoteles. To examine the structure and dynamics of the Earth's interior by monitoring the fine-structure of its gravity field.
- A second generation of geo-stationary METEOSAT satellites to provide continued and enhanced meteorological data.
- A series of polar orbiting satellites in support of ESA's Earth observation strategy which requires long-term global-datasets for monitoring the environment, and prudent management of the Earth's resources.

Foremost amongst the future ESA polar orbiting satellites is Envisat-1, to be launched in 1998. This satellite will carry the instruments listed in table 2.1 [2]. The ASAR instrument on Envisat-1 provides an enhanced form of the synthetic aperture radar sensors carried on ERS-1 and ERS-2 (which have a rapidly expanding user community [3]). The ASAR device delivers a flexible swath (able to deliver high or low resolution images) and exploits an alternating polarisation facility in its emitted signals for enhanced sensing of the returned signal.

The US's National Aeronautics and Space Administration (NASA) has undertaken a *Mission to Planet Earth*; a major program to examine global changes on Earth using a variety of spatial, spectral and temporal scales. The mission will be implemented by an Earth Observing System (EOS) which will be a suite of about 15 satellites to be launched over the next 20 years. These satellites are intended to provide a long-term database of changes, both natural and anthropogenic, in the Earth's environment. The EOS suite of satellites will have either a morning, descending node, equatorial crossing (optimised for minimal cloud cover observation of surface features) or an afternoon, ascending node, equatorial crossing (selected for its usefulness in meteorological observation). These two sets of EOS satellites are called the EOS-AM and EOS-PM satellites respectively.

The first EOS satellite, EOS-AM-1, will carry a payload of 5 sensors: the Multi-angle Imaging Spectroradiometer (MISR), an instrument for the Measurement of Pollution in the Troposphere (MOPITT), a sensor to monitor the Cloud and Earth's Radiant energy System (CERES), the Moderate Resolution Imaging Spectrometer (MODIS), and the Advanced Spaceborne Thermal Emission and Reflection Radiometer (ASTER).

Table 2.1. Instruments included in the Envisat-1 payload.

AATSR	Advanced Along-Track Scanning Radiometer
ASAR	Advanced Synthetic Aperture Radar
DORIS	Doppler Orbitography and Radio-positioning Integrated by Satellite
GOMOS	Global Ozone Monitoring by Occultation of Stars
LRR	Laser Retro Reflector
MERIS	Medium Resolution Imaging Spectrometer
MIPAS	Michelson Interferometer for Passive Atmospheric Sounding
MWR	Microwave Radiometer
RA-2	Radar Altimeter 2
SCIAMACHY	Scanning Imaging Absorption Spectrometer for Atmospheric Cartography
SCARAB	Scanner for Radiation Budget

The MERIS instrument on board Envisat-1 will complement the MODIS sensor included in NASA's EOS satellite payloads. MODIS, which will be carried on both the EOS-AM and EOS-PM spacecraft, has been designed to satisfy the following observational requirements, drawn from a need to understand the Earth as a complex integrated system, [4]:

– a wide field of view (2, 330 km);
– broad spectral coverage (36 bands in the range $0.4 - 14\mu$m);
– precise spectral band registration;
– high radiometric sensitivity;
– accurate calibration;
– variable spatial resolution (200m, 500m, 1000m at nadir);
– long operational life expectancy (providing 15 years of continuous data for the EOS satellite programme).

The ASTER device will have the highest spatial resolution of all of the EOS-AM-1 instruments [5]. The combination of high spatial resolution and wide spectral coverage should may help to reduce the problems encountered with *mixed* pixels: the reduction in pixel *footprint* reduces the likelihood that its *contents* are heterogeneous. Some of the key attributes of the ASTER sensor are listed below:

– 60 km field of view;

- broad spectral coverage including the visible, near infra-red, and thermal infra-red (14 bands in the range $0.5 - 11\mu m$);
- high spatial resolution (15m in the visible and near infra-red, falling to 30m in the thermal infra-red);
- stereo scopic capability in the along-track direction.

3. Investigation of Trends in the Processing Requirements of Remote Sensing Applications

Historically, the motivation for and principal civilian uses of remotely sensed data has been the need to gather information on a fairly gross scale for applications such as land-use, climatology and meteorology. For these sorts of applications the algorithmic complexity and computational power required to process the raw satellite imagery was formidable.

However, a pervasive trend in the requirements for remote sensing applications is the need for finer-grained analysis of particular regions and even urban areas. The reduction in the *granularity* of the data is not necessarily always accompanied by a reduction in the size of the area under scrutiny eg. the MARS (Monitoring of Agriculture by Remote Sensing) project carried out at the JRC in support of the European Union's Common Agricultural Policy requires detailed pan-European information about agricultural land-use on a field-by-field basis.

To some extent the trend towards more detailed information is driven by the increasing sophistication of satellite sensors and their ability to deliver higher spatial resolutions. In addition to this *driving force* the realm of remote sensing applications is also being *pulled* by the rapid development of technologies such as spatial databases and Geographical Information Systems (GISs). The need for readily available, timely, and cost-effective spatial and thematic data is a fundamental requirement of all GISs.

The potential market for such GIS support data is likely to be very large particularly where, through multi-temporal analysis of satellite imagery, the evolution of fundamental surface and sub-surface characteristics as well as land-use can be tracked through time.

This section of the paper will now examine certain case-study examples which support the above assertions about the trends in remote sensing applications.

Digital Terrain Mapping & Automated Product Generation. Allan [6] has proposed that a major obstacle preventing the wide-spread use of remotely sensed data is the spatial resolution that has until recently been available. Until the launch of the French SPOT satellites in the mid-80s non-military satellite sensors were designed for environmental and climatological monitoring. The data gathered by these systems was often intended for use on a continental scale. Allan [6] argues that such information while extremely important is of little use to utility companies, local and regional Governments, and other commercial companies whose need is for timely, accurate, and detailed information about specific localities.

Allan foresees a widening of the user-community and increased use of remotely sensed data as the new satellites with spatial resolutions of a few metres start to deliver their data to ground processing stations who can provide fast, customer orientated, imagery on demand via the Internet.

Gartner [7] shows how increased automation in the production of remote sensing products is a necessary trend if timely customer services are to be provided. The work of Gartner [7] tackles the difficult problem of extracting meteorological information from METEOSAT images. Fully automated meteorological product generation is still a future target but the trend is to reduce the amount of manual product rectification by using increasingly sophisticated automated tools and techniques.

Thompson and Mercer [8] have shown how the provision of timely remote sensing products can often be the only alternative if commercial companies are to retain their competitive edge. They cite the example of a European telecommunications company which, faced with stringent contractual time limitations, found that the only way to obtain detailed topographical information about parts of Indonesia was to use digital terrain models derived from SAR data obtained by satellites such as ERS-1 and RADARSAT.

Multi-Temporal Analysis. A recurring theme of recent initiatives to gather more data about the processes affecting the environment whether on a local or global scale is the increased need to carry out multi-temporal analyses of remotely sensed data. The aim of such multi-temporal analyses is to see how certain aspects of a study area change through time. The temporal resolution for such analyses can vary. For climatological studies it is expected that large quantities of data will be required spanning several decades of observations. For other studies monthly variations of particular attributes might be of interest but the temporal extent of the study might be restricted to one or two years.

Whether the temporal extent of a multi-temporal study is large or small the need to perform such multi-temporal analyses exacerbates the data processing bottleneck associated with remotely sensed data.

At the JRC studies are underway to estimate the rates of soil erosion in the arid climates of European countries which form the Mediterranean coast. The areas selected for detailed study in the Soil Erosion Model for Mediterranean Areas (SEMMED) are two parts of the island of Sicily. The study areas display characteristics that are typical of the problems encountered in the region: deforestation and increasing exploitation of the land for agriculture along with seasonal heavy rains which have the capacity for displacing large amounts of top-soil.

The data available for the JRC's SEMMED study consist of:

− 30 years of accumulated rainfall statistics;

- a soil and geological maps;
- hydrographic data;
- a digital terrain model;
- and a multi-temporal set of Landsat Thematic Mapper images.

The Landsat images span a temporal extent of approximately 10 years and consist of 7 images acquired in 1985 and six images acquired in corresponding seasons in 1993/1994.

Each of the Landsat images are *quarter images* of the Thematic Mapper's full-scene. Each image, includes data from all spectral channels and occupies 70 mega-bytes of storage. The satellite imagery is used to help determine important parameters in the SEMMED model: to determine the classes of the surface vegetation and a qualitative interpretation of the rate of soil erosion. The vegetation classes extracted from the imagery are incorporated into layers of a GIS and used to determine a quantitative assessment of the rate of soil degradation. Supervised and un-supervised techniques are used to analyse the imagery when identifying the vegetation types.

The Landsat images have to effectively be super-imposable, introducing problems of geometric rectification. The SEMMED analysis also requires the generation of a digital terrain model (DTM) to provide important elevation and slope data for estimates of run-off. The process of generating the DTM requires a large number of data points, extracted from the the Landsat imagery. The processing required to generate the DTM is considerable, up to 2 days of CPU supported by 160 mega-bytes computer memory, and the resulting model occupies 46 mega-bytes of disk storage.

The rainfall statistics available to the SEMMED model span 30 years and have been collected from more than 200 rainfall monitoring stations in the test areas. The SEMMED model however still needs to perform a computationally intensive interpolation to obtain the rainfall intensity data for those points in the test area not directly referenced by a monitoring station.

Monitoring the marine environment of the area off-shore of the Simeto study area in Sicily is another project in which the size of the acquired images presents practical problems. The aim of this study is to determine how the chlorophyll concentration and sedimentary deposition rates fluctuate with time. Here again, multi-temporal analysis of Landsat Thematic Mapper images is required. The high spatial resolution of the Thematic Mapper complements other remotely sensed data acquired using the Coastal Zone Colour Scanner (CZCS) sensor for monitoring ocean colour but which only delivers a resolution of 1km×1km. These datasets all pose familiar remote sensing data processing problems such as requiring atmospheric corrections and geometric rectification. The extra dimension arising from the need to perform multi-temporal analyses of these large data sets poses significant processing problems.

In the near future the MOMS-2 satellite will provide the project team with higher precision bathymetric data about the Sicilian sea-bed, eg. with a resolution of 5m×5m, which is required for elevation models which will allow detailed analysis of sedimentary processes taking place in estuarine and coastal plumes. These elevation models will provide input to geographical information systems which are also an important tool for the study.

The JRC also has a project to acquire forest information from remote sensing (FIRS) data. The FIRS project tries to assimilate information about the composition, condition, and evolution of European forests on a continental scale. The focus of attention for the FIRS project has been the mostly coniferous forests and taiga of the large forested belt of the temperate zones in the northern hemisphere: the Earth's tropical rain-forests having already been the subject of detailed in the Tropical Ecosystem Environment observation by Satellites (TREES) project carried out by the JRC and ESA.

Monitoring the temperate boreal forest involves large areas of the Earth's surface and does not require particularly high sensor resolutions (eg. spatial resolutions of the order of a kilometre are acceptable rather than tens of metres). Even so the quantity of data captured is still very large, due to the vast areas that are being surveyed.

The satellite imagery used in the study was gathered from two sources: the Landsat Thematic Mapper, and the NOAA satellites' AVHRR sensors. The high resolution Landsat images of selected test areas, in conjunction with ground-truth data, are used to develop important parameters in the FIRS model for estimating the biomass of the boreal forests. The models are then run using the lower resolution AVHRR data which covers the entire area of boreal forest being studied to provide estimates of the biomass present.

Increased Data Volumes - Through Increased Spectral Channels. The new satellite sensors display a trend towards acquiring data from finer spectral resolutions. The result is that the acquired data covers a larger number of discrete spectral channels. This trend will almost certainly improve the fidelity and quality of the final remote sensing products but increases the need to find fast and effective ways to fuse multi-spectral data during its analysis.

3.1 Commercialisation of Remote Sensing Data

In tandem with the rapidly evolving technical specifications of satellite hardware, another important development in the field of RS has taken place.

The over-arching national, continental, and global importance of the information present in remotely sensed data, along with the enormous initial cost of hardware development and deployment has meant that the first 20 years of RS has been dominated by state-run national and international space centres.

For example, from 1972 to 1982 the US's Landsat program was administered by NASA and until late 1985 by the National Oceanic and Atmospheric Administration (NOAA). However in 1984, the US Congress passed the Landsat Remote Sensing Commercialisation Act which authorised the moving of the Landsat program into the private sector. Since late 1985, the Earth Observation Satellite Company (EOSAT) has administered the Landsat program as a commercially viable company. A key component of this activity has been the need to broaden the Landsat user base. Part of EOSAT's mission is to become a one-stop centre able to provide clients with diverse, high quality, timely and cost-effective products.

The trend towards increasing commercialisation has also extended to Europe. The almost annual reductions in funding have made it imperative for organisations such as the ESA to find commercial partners and customers to help fund satellite launches and reduce the cost to member states of the derived RS data products.

Over recent years an enduring trend has emerged in the market for remote sensing data products. The potential applications of RS data are so diverse that it is very difficult for products derived at satellite ground-stations to satisfy the particular requirements of diverse customers whose interests are very specifically targeted. Rather, the ground-stations need to concentrate upon fundamental operations on the raw data such as geometric-rectification, major feature extraction and classification to produce a *generic* partially processed product. This *generic* data then needs to be quickly despatched to particular customers who need tools able to operate upon the *generic* image in near real-time to extract the particular information they require. This trend towards empowering the end-user to carry out their own data processing locally is akin to the client-server computer system architectures which has proved such a milestone in the development of the *Internet*.

4. Evaluation of Current and Potential Uses of NN-HPCN Systems in Remote Sensing

A recurring theme in much of the reported work on NNs is the extended training times for the networks. A neural network consists of a large collection of very simple processing units - the *neurons*. The processing units exchange only very simple messages - typically consisting of the magnitude of each unit's activation. These simple messages are transmitted through weighted connections - the *synapses*. Neural network training consists of a slow, iterative gradient-descent to find a pattern of weighted connections which allow the network to accomplish its training task. The determination of an effective pattern of network connectivity is the processing bottleneck when using neural networks.

Table 4.1 gives some examples of the computation-time problems that must be overcome when training neural networks. Every network listed in table 4.1 is a Multi-Layer Perceptron (MLP), of the sort that is widely used in remote sensing to obtain thematic classifications. Each MLP has 3 layers of *neurons*: an input layer which receives the satellite image data; a single hidden layer; an output layer which shows the classification that the network has determined for each pixel of input data. Successive network layers are connected to each other by a matrix of connection-weights. Table 4.1 is in two halves, the first half shows several MLPs which might be used to classify data provided by the Landsat Thematic Mapper: using 6 of the Thematic Mapper's channels as input and classifying each input into one of 6 thematic classes. The difference between each MLP is the size of the hidden layer. For technical reasons, which are beyond the scope of this paper, the hidden layer size is the only variable that can be changed without fundamentally affecting the overall classification problem.

The second half of table 4.1 contains MLPs that might be used to classify data acquired by new satellite sensors such as MODIS. In these cases the MLP input layers accept data from 30 of the sensor's channels and could be used to classify each input pixel into one of 12 thematic classes. Again, the only change between each these networks is the size of the hidden layer. Every one of the MLPs shown was trained on an input dataset of 2000 pixels and for 600 *epochs* (each epoch of training involves a complete pass through the training dataset and a subsequent update of the MLP's weights). A period of 600 epochs of training was selected as a reasonable sample from which the average computation-time per epoch could be determined: none of the MLPs were expected to successfully complete their classification task in this time and none did so. Every MLP was trained on the same computer, a SUN SPARC ULTRA with 80 mega-bytes of RAM, and with the input dataset stored locally to avoid any undesirable delays due to network file-access to the training dataset. For each MLP, table 4.1 shows the average elapsed time per epoch of training and, derived from that figure, the average elapsed time per training pattern in the input dataset.

Table 4.1. Some example remote sensing MLP training times.

MLP Configuration (I:H:O)	Elapsed Time per Epoch (Seconds)	Elapsed Time per Pattern (Milli-Seconds)
6:4:6	0.22	0.11
6:8:6	0.36	0.18
6:12:6	0.49	0.245
6:16:6	0.63	0.315
6:20:6	0.76	0.38
30:20:12	2.31	1.16
30:40:12	4.91	2.46
30:60:12	6.93	3.47
30:80:12	9.38	4.69
30:100:12	14.33	7.17

Many of the parameters which effect neural network training are assigned values on a heuristic basis. In [9] heuristic guidance is given about the size of training dataset if the resulting 3-layer MLP network is to have good generalisation performance with novel data. If ϵ is the fractional error that must be satisfied by at least half of the training patterns at the MLP output layer and W is the number of network weights & biases that must be updated at each epoch, then the number of training patterns required for good generalisation performance (S) can be estimated using the following inequality

$$S > \frac{W}{\epsilon}.$$

The fully-connected structure of the MLP architecture, in which each node in a layer is connected to **every** node in the succeeding layer of the network, means that increasing the number of nodes in a layer can have a dramatic impact upon the value of W. For any MLP network the value of W is given by

$$W = \sum_{i=2}^{n} (U_{i-1} \times U_i) + U_i$$

where i indexes each layer, n is the total number of layers and U is the number of units in the specified layer. Using the above inequality for the 6:16:6 and 30:80:12 networks of table 4.1 and assuming a fractional MLP training tolerance of 0.1, good training datasets would require $2,140$ pixels and $34,520$ pixels respectively. Using the appropriate figure from the third column of table 4.1 (t) allows the time taken to train each network to be roughly estimated by

$$T = S \times t \times E$$

where E is the number of epochs of training. Thus the time taken (in hh:mm:ss) to train these two MLP networks for 600 epochs if reasonable generalisation performance is required is 00:06:44 and 26:58:02 respectively.

Clearly, the training time of the 30-channel MLP is far longer than that which would be acceptable for any near real-time operational classification system. Importantly, the above estimates for the MLP training times assume (purely to allow easy comparison of the effect of increased network size) that both MLPs satisfy their error criteria in 600 epochs of training. However it is highly likely that for larger networks, being trained to perform more complex classifications, the number of epochs required for acceptable classification performance could be much longer.

However, the very simple processing of each unit in a neural network and the repetitive nature of the processing to derive the connection weights, means that neural network training lends itself to a parallel processing implementation.

Figure 4.1 is a schematic diagram of the Siemens SYNAPSE-1 neurocomputer. Neural networks justify their own form of parallel processing device (neurocomputers) because their structure and processes are so well suited to a parallel implementation. Unlike attempts to parallelise conventional algorithms - characterised by the need for a centralised executive which usually performs algorithmic tasks sequentially, often requiring complex problem decomposition into parcels that can be performed in parallel - neural network training algorithms and operational characteristics lend themselves much more readily to neuron/unit level concurrency. The SYNAPSE-1 neu-

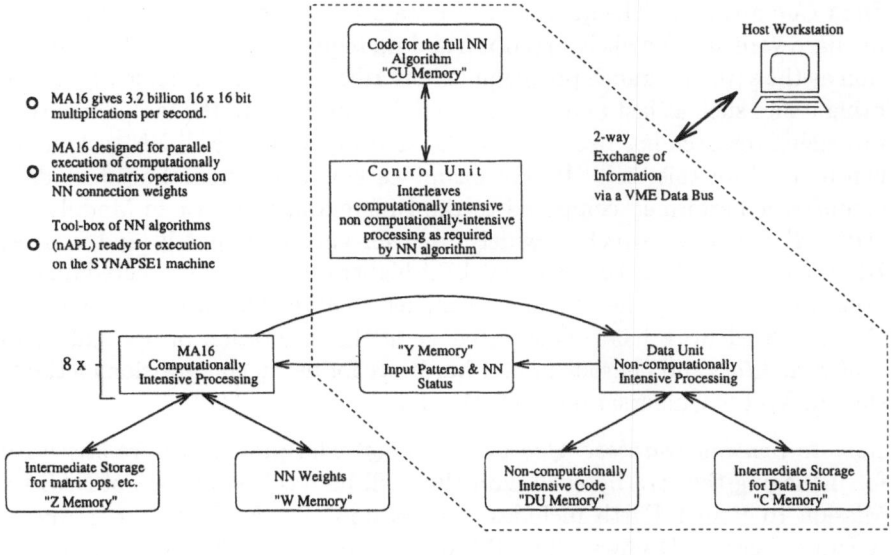

SYNAPSE1

(SYnthesis of Neural Algorithms on a Parallel Systolic Engine 1)

Fig. 4.1. An overview of the SYNAPSE-1 Neuro-computer.

rocomputer has been designed to specifically perform the parallel operations of neural networks and should avoid the overheads in terms of problem decomposition, synchronisation and inter-process communication which blights attempts to parallelise conventional algorithms. Highlights of SYNAPSE-1's features include:

- special hardware to perform the matrix-operations upon the NN weights (which is at the heart of all NN training);
- very fast access to memory which is dedicated to storing the very large numbers of weights used in large networks - such as those found in remote sensing;
- separate processors for the less computationally intensive tasks such as buffering the NN training and testing patterns.

With potential hardware solutions, such as SYNAPSE-1, to the problem of neural network training the attractiveness of neural networks in remote sensing, where the training data sets are very large, is greatly enhanced.

The need for innovative techniques, such as neural networks, is now becoming more widely recognised in the remote sensing user community. Some important application areas are described in the remainder of this section of the paper.

Data Compression. The quantities of data produced by current remote sensing hardware and the likely explosion of data requiring storage and manipulation as the satellite launch programmes described in section 2 come to fruition bring issues such as fast and effective data compression to the fore. Some very promising research has been carried out to compare how well NN techniques can be used for this task. [10] has examined how well a neural network data compressor performed compared to the Differential Pulse Code Modulation (DPCM) technique which is widely used for encoding speech, and images. His results show that the NN can deliver higher quality reconstructed images even after encoding the image at a higher compression ratio than DPCM. Other related work has concentrated upon the generation of optimal *codebook generators* and neurocomputer hardware for use in compression methods such as Vector Quantisation (VQ) [11, 12].

Remote Sensing Database Query Processing. The proposed infra-structure for managing the tera-bytes of data that will be received as part of NASA's Mission to Planet Earth includes the creation of eight Distributed Active Archive Centres (DAACs). The DAACs will be responsible for storing and allowing access to the vast amounts of satellite imagery - much of which will have to be highly compressed for long-term storage. In order for this archive of information to be useful, fast and efficient methods have to be found by which diverse remote-sensing users can identify relevant imagery for their applications. Such database queries are likely to place a heavy computational burden upon the DAACs requiring high-performance computing in support of query processing which is at odds with the DAACs primary archiving activities. One possible solution to this problem [13] is to train neural network classifiers on a high performance computer that is remote from the DAAC and then *send* the trained classifier to the DAAC where it quickly extracts the desired imagery from the compressed DAAC archive. The still-compressed imagery can the be efficiently despatched to the originator of the query. Such a strategy has the advantage of reducing the computational burden on the DAAC and the amount of satellite imagery traffic travelling on the Internet.

Structural Manipulation. The occurrence of *mixed pixels* is a common problem with remotely sensed data that can hinder its reliability and usefulness for applications which require high accuracy. Indeed, it can often be very difficult to determine the accuracy/correctness of any system's performance when the classification of a mixed pixel is required. It may be possible to use certain neural network architectures to recognise commonly occuring mixed-pixel compositions from their spectral signatures. These neural network architectures through their ability to extract recurring structural forms, may also be of use for GIS applications of remotely sensed data where the ability to perform *map generalisation* of particular objects rendered at different scales is an active area of research. The ability to recognise recurring structural forms may also be important in the analysis wind fields detected in METEOSAT imagery. Staff at the European Space Agency (ESA) face an important problem when attempting to dis-ambiguate competing solutions to complex numerical models [14]. Attempts to emulate human performance on the dis-ambiguation task has so far proved unsatisfactory. A speculative hypothesis is that using structure recognising NNs could provide the basis for an improved automated dis-ambiguation system.

Atmospheric Corrections. Making corrections for the effect of atmospheric distortions upon satellite imagery may be a useful application of neural networks' ability to apply a complex data transformation which degrades gracefully. Here, the transformation might be more pronounced for the off-nadir pixels with the transformation becoming minimal for pixels at nadir.

Geometric Rectification. Liu and Wilkinson [15] have shown that errors which arise in satellite imagery because of fluctuations in satellite orbit, the rotation of the Earth, and the relief of the terrain can be rectified using NNs. The work of Liu and Wilkinson [15] also shows that training times for the rectifying neural networks can be extensive. However, the deployment of HPCN systems may alleviate the problems of extended training times. If so, then the inherent parallelism of the trained network will deliver significant improvements over conventional rectification algorithms - particularly when the image datasets are very large.

Feature Extraction and Detection. Neural networks have a proven ability to perform as well as more conventional techniques for extracting and detecting particular features from satellite imagery. Of particular interest is the need to identify discrete point, linear and areal features within an image. The identification of areal features is closely related to the the problem of image segmentation. The standard algorithmic solutions to these problems are hard to automate since they rely heavily upon an expert's selection of parameters. The parameter settings themselves can vary widely from image to image. Here the property of neural networks that allows them to act as general purpose, non-parametric, tools for extracting information from satellite data is a distinct advantage.

Stereo Matching. Uses a high quality stereoscopic pair of images of a target area and then extracts three-dimensional information to provide, for example, digital elevation information directly from remotely sensed data. Standard, computationally very intensive, algorithms are now available for this sort of process. Neural network solutions, such as that demonstrated by [16], offer the potential benefits of parallelising the whole process but also the possibility of enhancing the performance of other operations such as feature detection/extraction.

SAR Interferometry. This is a relatively new technique [17] which exploits the coherence RADAR signals obtained from separate SAR images of the same area to develop interference patterns with convey detailed information about the observed area. This form of *synthetic interferometry* can be applied to SAR data acquired by satellites such as ERS-1, ERS-2 and RADARSAT. However, several problems remain with the use of SAR interferometry. The SAR data is nearly always accompanied by noisy speckle effects, requiring the use of speckle-filtering techniques. The interferograms which are the output of the process also frequently contain interference-fringe discontinuities. The automated *closure* of such discontinuities is an important potential application area for NNs.

Parallel Processing. Tests at the JRC [18] have shown that the time taken to perform certain types of image-noise reduction, using an iterative *moving window* technique, can be significantly reduced by using parallel processing. The parallelism tested, involved splitting the filtering problem between a small network of workstations. Essentially this sort of parallelism relies upon a decomposition of the problem between sequential machines. A three-fold speed-up of the overall filtering process was observed. The parallelism described in [18] does not rely upon the use of massive parallelism such as that offered by the Siemens SYNAPSE-1 [19] machine. Nor does the speed-up result from the parallelism offered by NNs. Combining the inherent parallelism of NNs and parallel hardware seems likely to make an additional contribution towards deriving timely high-quality products from remotely sensed data: beyond those obtained by the method of problem decomposition as described by [18].

Further support for allying neural network techniques with the parallelism of high-performance neurocomputer technology arises, paradoxically, from a weakness of neural networks used in the domain of land-cover classification. Research into the portability of neural network classifiers trained using remotely sensed data has shown that good classification performance is very geographically specific [20]. Classification performance of the order of 70% can readily be obtained when testing the trained networks on new data drawn from the same locality as the training data. However performance falls to about 13% when the same networks are tested on data drawn from a geographically remote region. Such results emphasise the need for neural network classifiers that can be quickly and easily trained for use by users in diverse geographic locations and application domains - a task for which neurocomputers such as the SYNAPSE1 should be well-suited.

Thus far the use of neural network techniques in remote sensing applications has been largely focussed upon use of the multi-layer perception (MLP) architecture [21]. This architecture is also very widely used beyond remote sensing because of its generality and ease of use.

However, the pre-eminence of the MLP architecture should not hinder further research to explore the large array of alternative neural network architectures which may have particular strengths in certain remote sensing applications. Eg. Counter-propagation Networks, Learning Vector Quantisation, Kohonen Networks, CMAC, Functional-link Networks, Fuzzy ART-MAP Networks, may all find particular remote sensing tasks to which they are well-suited.

5. Conclusion

The use of the Multi-Layer Perceptron (MLP) neural network architecture has been prevalent in remote sensing neural network research to-date. This is not surprising since the MLP is one of the simplest neural network architectures and it is the most popular choice for neural network research in other domains.

However, the MLP architecture, despite its generality of use, requires an extended training phase (due to its fully connected weights) - a problem which is exacerbated in remote sensing where increasingly large training data-sets are the norm.

The indications are that neural networks can play an increasingly important role in tackling the problems of the remote sensing community. They find straightforward parallel-processing implementations which should allow them to be trained much faster and also, once trained, to be deployed to provide high quality, timely, and reliable products.

However, widespread use of neural networks within the remote sensing user communities will depend upon hard evidence that:

– they can carry out the operations and transformations required;
– in tandem with highly parallel hardware they are able to generate high quality products much faster than by conventional means - ultimately in near real-time.

Attempting to gather evidence about these two pivotal issues is the motivation behind the sort of bench-marking exercise, of which this paper has been the first stage.

References

1. W. Bechtel and A. Abrahamsen, *Connectionism and the Mind: an introduction to parallel processing in networks*, Cambridge, MA: Blackwell, 1991.

2. C. J. Readings and P. A. Dubock, "Envisat-1: Europe's Major Contribution to Earth Observation for the Late Nineties", *European Space Agency Bulletin*, no. 76, pp. 15–28, 1993.
3. S. Karnevi, E. Dean, D. J. Q. Carter, and S. S. Hartley, "Envisat's Advanced Synthetic Aperture Radar: ASAR", *European Space Agency Bulletin*, no. 76, pp. 30–35, 1993.
4. T. S. Pagano and R. M. Durham, "Moderate Resolution Imaging Spectroradiometer (MODIS)", in *Sensor Systems for the Early Earth Observing System Platforms*, pp. 2–17, SPIE - The International Society for Optical Engineering, Washington USA, 1993.
5. Y. Yamaguchi, H. Tsu, and H. Fujisada, "Scientific basis of ASTER instrument design", in *Sensor Systems for the Early Earth Observing System Platforms*, pp. 150–160, SPIE - The International Society for Optical Engineering, Washington USA, 1993.
6. J. Allan, "A new era for remote sensing and GIS", *GIS Europe*, vol. 5, no. 4, pp. 24–25, 1996.
7. V. Gartner, "Towards automatic product generation from METEOSAT Images", *European Space Agency Bulletin*, Number 8, 1994,
   http://esapub.esrin.esa.it/pointtobullet/gart82.htm.
8. M. D. Thompson and J. B. Mercer, "Digital Terrain Models from RADARSAT", *Earth Observation Magazine Online*, March 1996:
9. G. Simpson and K. Li, "Artificial Neural Networks: Solutions to Problems in Remote Sensing", Technical Report EOS-92/000(16000)-RP-001, Earth Observation Sciences (EOS), Farnham, Surrey, 1992.
10. C. N. Manikopoulos, "Neural network approach to DCPM system design for image coding", *IEE Proceedings-I: Communication, Speech and Vision*, vol. 139, no. 5, pp. 501–507, 1992.
11. M. Lech and Y. Hua, "Image vector quantization using neural networks and simulated annealing", in *Proceedings of the International Conference of Image Processing and Applications*, pp. 534–537, Maastrich, the Netherlands, 1992.
12. W. C. Fang and et al, "A VLSI neural processor for image data compression using self-organization networks", *IEEE Transactions on Neural Networks*, vol. 3, no. 3, pp. 506–518, 1992.
13. R. A. Schowengerdt and J. D. Paola, "Parallel computing and data compression for pattern matching in remote sensing image databases", in *EUROPTO Rome 1994*, Society of Photo-Optical Instrumentation Engineers, Washington, 1994.
14. AEA Technology, "Study on the applicability of neural networks to significant event recognition: Wind Vector Extraction Project", Technical Report SWD-6201-FR-AT, ESA/ESOC Contract No: 10119/92/D/IM, 1994.
15. Z. K. Liu and G. G. Wilkinson, "A neural network approach to geometrical rectification of remotely sensed imagery", Technical Report I.92.118, European Cimmission Joint Research Centre, 1992.
16. G. Loung and Z. Tan, "Stereo matching using artificial neural networks", *International Archives of Photogrammetry and Remote Sensing*, vol. 29, no. B3, pp. 417–421, 1992.
17. S. N. Coulson, "SAR Interferometry with ERS", *Earth Space Review*, vol. 5, no. 6, pp. 9–16, 1996.
18. M. J. Mineter, "Final Report: Pilot Study for a Parallel Processing Approach to Raster Image Generalisation with Potential Application to Forest Mapping", Technical Report 1-March-96, EC Joint Research Centre, Ispra Italy, 1996.
19. Siemens Nixdorf Informationssysteme AG, "SYNAPSE1 N110 Technical Description", 1995. Document No: S26611-K1-Z1-01-7618.

20. DIBE - University of Genoa, "Portability of Neural Classifiers for Large Area Land Cover Mapping by Remote Sensing", 1995. Report produced under contract to the European Commission's Joint Research Centre, Ispra.
21. D. E. Rumelhart, J. L. McClelland, and The PDP Research Group, *Parallel Distributed Processing: Explorations in the Microstructure of Cognition*, vol. 1. Cambridge, Massachusetts: MIT Press, 1986.

General Discussion

Graeme G. Wilkinson

School of Computer Science and Electronic Systems,Kingston University,
Penrhyn Road, Kingston Upon Thames, Surrey KT1 2EE, UK.

The participants to the COMPARES workshop were asked to consider and reflect upon a set of key questions which relate to the future use of neurocomputing in remote sensing. These questions were as follows:-

i. Why use neural networks in the analysis and interpretation of remote sensing data ? Do they really offer advantages ? Are they really special ?

ii. Classification has been the main application to date but has an impasse been reached ?

iii. Is it possible or necessary to build a "pan-European-classifier" based on a giant modular neural network ?

iv. How can neural network systems be made totally user-friendly ?

v. Is special purpose hardware needed to operationally exploit neural networks in remote sensing ?

vi. Should new / less common network models / architectures be explored ?

vii. Are there any novel applications of neural networks in remote sensing ?

viii. What future level of investment on R&D on neural networks in remote sensing is justified ?

Some of these issues were discussed openly and in small private groups throughout the workshop and have also been addressed in the individual articles in this volume. However a few participants of the workshop produced written comments during or shortly after the meeting on these issues. Their comments, in a paraphrased and abbreviated form, are given below:

Comments from Dr. A. Nielsen (Technical University, Denmark)

Overall neural networks appear to offer marginal benefit compared to more conventional methodologies. They are expensive computationally compared to methods that perform similarly. For image classification the key issue is the production of "quality" training data and quality features rather than the use of marginally different discriminators. Special purpose hardware is probably needed in order to make best of neural networks, but there is no need to produce a pan-Euro classifier. For the future it is not necessary to devote funds to research on neural networks in remote sensing -there is no need to look at application to remote sensing as a special case.

Comments from Mr. C. Stephanidis (University of Dundee, UK)

The central problem at present in remote sensing seems to be that too little is known about the nature of the new forms of satellite data such as synthetic aperture radar data and imaging spectrometry data. Also, a "global" approach is needed using contextual information and multi-source data -though it remains extremely difficult to obtain data from different satellite sensors. The key question is now how to get closer to user requirements without using complex tools. Furthermore, in order to compare data analysis methodologies, it is necessary to develop a database of standard test data sets that could be used to benchmark different approaches.

Comments from Dr. T. Schouten (Katholieke Universiteit Nijmegen, Netherlands)

Neural networks really do offer advantages compared to more traditional approaches especially in terms of speed, learning and generalisation. They do have special properties that offer significant benefits, though current use is restricted to solving relatively simple statistical analysis problems. The classification problem is not yet solved, a better definition of the problem and better data sets are required. A pan-European classifier could be useful though would be premature at present. Initially better data sets are needed to aid construction and testing of such a classifier. It is not necessary to use special purpose hardware for neural networks as general purpose parallel machines are sufficient. In the future methodological improvements could be made by combining statistical and logical processing with neural networks. There is future scope for using neural networks to solve inverse problems in remote sensing and to make more use of image texture information in land use analysis.

Comments from Dr. F. Roli (University of Cagliari, Sardinia, Italy)

Artificial neural networks are still worthy of attention since the biological neural networks seem to work so well. They don't seem to offer particular advantages at present only complementary characteristics with respect to traditional statistical classifiers. They could become very special in the future, though this depends on hardware and software developments. In order to make progress in classification research it is necessary to develop some standard benchmark datasets as has been done by the US Mail service for hand-writing recognition. The preparation of such benchmarks should have priority over building a pan-European classifier as it is first necessary to be able to compare results from different approaches in a reliable way. Special purpose hardware is definitely needed to really exploit neural networks. It could be useful to examine the potential of new and less common neural network models, though existing neural networks and statistical methods offer a good complementarity.

In the future, one of the most important new applications of neural networks in remote sensing could be the analysis of hyper-spectral data. Future research funding is necessary for work on neural networks in remote sensing, though this could be seen as basic research which appears not to have a priority at European Union level.

Comments from Professor G. Foody (University of Salford, UK)

Neural networks do offer advantages in remote sensing over other more conventional methods. The end of the road has definitely not been reached in classification research -but the emphasis should be on "mapping" rather than "classification". The word classification is being used as equating to mapping - and this is not accurate. Remotely sensed data can only be used for three applications - mapping, monitoring and parameter estimation. Mapping is the "base" application and generally classification is used as the tool to achieve the mapping. It is probable that progress is actually being made in mapping but the tools used don't show this. More mapping research is definitely needed and neural networks are appropriate techniques to use. The accuracy of the mapping, however, cannot be assessed by the conventional measures (e.g. kappa, percent correct, tau etc.) and so it can not be argued that progress is not being made if the yardstick used to evaluate progress is flawed. Another important point to raise, however, is that unrealistically high standards may be being set - certainly other forms of map data are markedly less accurate but happily accepted and used.

The construction of a large pan-European classifier may be sensible, though this should be better thought of as a general neural network with extra inputs for latitude, solar zenith angle, landscape heterogeneity etc. Neural systems should be sufficiently user-friendly for general use, and a key requirement is to help the user choose a good network architecture. The use of special purpose neural hardware may be justified by some applications though this would prevent the method being widely applied. New and less common neural network architectures should be explored especially if it is desirable for relatively un-trained users to carry out data analysis. It may be useful to explore neural networks that largely build themselves. There are probably many more applications of neural networks in remote sensing other than classification.

Comments from Professor P. Mather (University of Nottingham, UK)

Neural networks are currently "fashionable" in remote sensing, though they seem not to be necessarily more effective than other methods and are not easy to design and to train. Their main advantage lies in combination of data without parametric requirements. The principal limitations to classification at present are not technique related.

The important issue is internal and external consistency of the data. Much more effort should be devoted to use of other spatial data such as geology, soil type, climate and topography in image analysis. This should be the focus of the possible pan-European classifier. The task of making neural networks more user-friendly should be left to software vendors. There should be a link-up with artificial intelligence and interface design. Special purpose neural hardware may be needed, though more efficient software may be a better approach.

It is also important to note that some unsupervised data analysis methods are several orders of magnitude more tedious computationally than supervised statistical classification. New neural architectures and models should be explored in remote sensing as that is how science develops. New applications of neural networks in remote sensing should be examined such as geometrical correction and multi-sensor data fusion. Neural network research should not dominate in receiving funding for methodological research in remote sensing as there are many other important methodological issues.